Effect of Pathogen Load on Pathogen Removal by Conventional Treatment

Awwa Research Foundation

About the Awwa Research Foundation

The Awwa Research Foundation (AwwaRF) is a member-supported, international, nonprofit organization that sponsors research to enable water utilities, public health agencies, and other professionals to provide safe and affordable drinking water to consumers.

The Foundation's mission is to advance the science of water to improve the quality of life. To achieve this mission, the Foundation sponsors studies on all aspects of drinking water, including supply and resources, treatment, monitoring and analysis, distribution, management, and health effects. Funding for research is provided primarily by subscription payments from approximately 1,000 utilities, consulting firms, and manufacturers in North America and abroad. Additional funding comes from collaborative partnerships with other national and international organizations, allowing for resources to be leveraged, expertise to be shared, and broad-based knowledge to be developed and disseminated. Government funding serves as a third source of research dollars.

From its headquarters in Denver, Colorado, the Foundation's staff directs and supports the efforts of more than 800 volunteers who serve on the board of trustees and various committees. These volunteers represent many facets of the water industry, and contribute their expertise to select and monitor research studies that benefit the entire drinking water community.

The results of research are disseminated through a number of channels, including reports, the Web site, conferences, and periodicals.

For subscribers, the Foundation serves as a cooperative program in which water suppliers unite to pool their resources. By applying Foundation research findings, these water suppliers can save substantial costs and stay on the leading edge of drinking water science and technology. Since its inception, AwwaRF has supplied the water community with more than $300 million in applied research.

More information about the Foundation and how to become a subscriber is available on the Web at **www.awwarf.org.**

Effect of Pathogen Load on Pathogen Removal by Conventional Treatment

Prepared by:
Prapakorn Assavasilavasukul and **Gregory W. Harrington**
Department of Civil and Environmental Engineering
University of Wisconsin – Madison, 1415 Engineering Drive, Madison, Wisconsin 53706

Boris L. T. Lau
Department of Civil and Environmental Engineering
Duke University, Box 90287 Hudson Hall, Durham, NC 27708

Rebecca M. Hoffman
Wisconsin State Laboratory of Hygiene
University of Wisconsin – Madison, 2601 Agriculture Drive, Madison, Wisconsin 53707

and

Mark A. Borchardt
Environmental Microbiology Laboratory
Marshfield Clinic Research Foundation, 1000 North Oak Avenue, Marshfield, Wisconsin 54449

Sponsored by:
Awwa Research Foundation
6666 West Quincy Avenue, Denver, CO 80235-3098

Published by:

Awwa Research Foundation

Distributed by:

American Water Works Association

IWA Publishing

DISCLAIMER

This study was funded by the Awwa Research Foundation (AwwaRF). AwwaRF assumes no responsibility for the content of the research study reported in this publication or for the opinions or statements of fact expressed in the report. The mention of trade names for commercial products does not represent or imply the approval or endorsement of AwwaRF. This report is presented solely for informational purposes.

ISBN 978-1-84339-264-4

CONTENTS

TABLES

FIGURES

FOREWORD

The Awwa Research Foundation is a nonprofit corporation that is dedicated to the implementation of a research effort to help utilities respond to regulatory requirements and traditional high-priority concerns of the industry. The research agenda is developed through a process of consultation with subscribers and drinking water professionals. Under the umbrella of a Strategic Research Plan, the Research Advisory Council prioritizes the suggested projects based upon current and future needs, applicability, and past work; the recommendations are forwarded to the Board of Trustees for final selection. The foundation also sponsors research projects through the unsolicited proposal process; the Collaborative Research, Research Applications, and Tailored Collaboration programs; and various joint research efforts with organizations such as the U.S. Environmental Protection Agency, the U.S. Bureau of Reclamation, and the Association of California Water Agencies.

This publication is a result of one of these sponsored studies, and it is hoped that its findings will be applied in communities throughout the world. The following report serves not only as a means of communicating the results of the water industry's centralized research program but also as a tool to enlist the further support of non member utilities and individuals.

Projects are managed closely from their inception to the final report by the foundation's staff and large cadre of volunteers who willingly contribute their time and expertise. The foundation serves a planning and management function and awards contracts to other institutions such as water utilities, universities, and engineering firms. The funding for this research effort comes primarily from the Subscription Program, through which water utilities subscribe to the research program and make an annual payment proportionate to the volume of water they deliver and consultants and manufacturers subscribe based on their annual billings. The program offers a cost-effective and fair method for funding research in the public interest.

A broad spectrum of water supply issues is addressed by the foundation's research agenda: resources, treatment and operations, distribution and storage, water quality and analysis, toxicology, economics, and management. The ultimate purpose of the coordinated effort is to assist water suppliers to provide the highest possible quality of water economically and reliably. The true benefits are realized when the results are implemented at the utility level. The foundation's trustees are pleased to offer this publication as a contribution toward that end.

David E. Rager
Chair, Board of Trustees
Awwa Research Foundation

Robert C. Renner, P.E.
Executive Director
Awwa Research Foundation

ACKNOWLEDGMENTS

The authors of this report wish to thank numerous individuals and organizations for their contribution to this project. Continuous flow centrifuges were donated to the University of Wisconsin by the United States Environmental Protection Agency and by the Milwaukee Department of Health. Centrifuge belts were meticulously prepared by Lynette Makowski and donated to the project by the Gambro Corporation. Microbiological counts were performed by Marty Collins of the Wisconsin State Laboratory of Hygiene and by Susan Spencer of the Marshfield Clinic Research Foundation. Numerous University of Wisconsin students participated in pilot plant operations throughout the project. These included Jian Yang, Kala Fleming, Steven Leung, Eric Phillippi, Eric Habermeyer, Eric Glisch, Katie Hassing, Adrienne Kuehl, Ben Reed, Brandon Herbert, Shawn Hall, John Ells, Tyler Rongstad, Anthony Heddlesten, Fahrol Jaafar and Tim Dallmann.

The authors also thank My-Linh Nguyen and John Albert, the AwwaRF project managers, and members of the project advisory committee for their support and suggestions throughout the project. Members of the project advisory committee were Frank Schaefer of the US Environmental Protection Agency, Joan Oppenheimer of Montgomery Watson Harza, and Mark Jenkins of the US Department of Agriculture.

ACKNOWLEDGEMENTS

EXECUTIVE SUMMARY

INTRODUCTION AND OBJECTIVES

Several studies have evaluated the removal of *Cryptosporidium* oocysts, *Giardia* cysts, and other emerging waterborne pathogens in conventional drinking water treatment. For example, Harrington *et al.* (2001) observed 2.1 ± 0.2 log removal of *Cryptosporidium parvum* oocysts while Nieminski and Ongerth (1995) reported 3.0 ± 0.6 and 3.4 ± 0.7 log removal of *Cryptosporidium* oocyst and *Giardia* cysts, respectively. Based on these and other studies (e.g., Patania *et al.* 1995, McTigue *et al.* 1998, Emelko *et al.* 2000, Huck *et al.* 2001), the USEPA has established that conventional treatment achieves an average 3-log removal of *Cryptosporidium* oocysts (USEPA 2003). As a result, when the second stage of the Long-Term Enhanced Surface Water Treatment Rule was finalized in early 2006 (USEPA 2006), conventional treatment plants received a 3-log credit for oocyst removal if they met the filtration requirements of the rule. Spike concentrations used in these experiments are 6 to 8 orders of magnitude higher than those observed in practice. Occurrence surveys have shown that the geometric means for *Cryptosporidium* oocyst concentrations in raw surface waters are around <0.001 to 3 oocysts/L, with a highest observed concentration of 484 oocysts/L (LeChevallier *et al.* 1991, LeChevallier and Norton 1995, States *et al.* 1997, LeChevallier *et al.* 2003). Thus, there is considerable uncertainty about pathogen removals that would be attained under more realistic situations.

The overall goal of this study was to evaluate pathogen removals in conventional water treatment at pathogen concentrations nearer to those found in the aquatic environment. The specific objectives of this study were to:

1. Measure the removal efficiencies of *Cryptosporidium* oocysts and *Giardia* cysts that have been seeded at more realistic concentrations of 10^1 pathogens/L and describe removal efficiency as a function of seeded pathogen concentration over the range of 10^1 to 10^5 pathogens/L.

2. Evaluate the differences between grab and composite sampling in quantifying pathogen removals.

3. Determine whether sorting by flow cytometry prior to microscopic enumeration is an alternative way of detecting pathogens in natural waters.

METHODS

This project evaluated pathogen removal with a unique combination of experimental and concentration methods. In order to characterize the removal of waterborne pathogens, experiments were conducted with different initial pathogen concentrations and sampling methods at the University of Wisconsin pilot plant. Grab and continuous sampling methods were used. Grab sampling is useful for observing filter effluent quality at a particular instant while continuous sampling indicates the average performance of the treatment process over an extended time period. To minimize the hindrance of raw water quality as a confounding variable, treatment trains were run in parallel to facilitate better comparison of treatment performance between trains due to the effect of pathogen spike dose. Continuous separation channel centrifugation was used to concentrate pathogens while immunofluorescence microscopy was used for pathogen enumeration. Particle sorting with flow cytometry prior to microscopic enumeration (FC sorting-microscopy) was compared with the more conventional method of microscopic enumeration.

For all experiments, the pilot plant received raw water from Lake Mendota in Madison, Wisconsin. The raw water had a moderate to high dissolved organic carbon (DOC) concentration of 3 to 6 mg/L, a low specific UV absorbance (SUVA) of 1 to 2 L/mg·m, a high alkalinity of 250 to 330 mg/L as $CaCO_3$, a relatively low turbidity of 2 to 5 ntu, and a relatively low total particle count of 1000/mL to 9000/mL.

CONCLUSIONS

1. The removal of *Cryptosporidium* oocysts and *Giardia* cysts through conventional treatment was observed to depend on initial pathogen spike dose. Removal increased with increasing spike dose and statistical analyses suggested that log removal differences between low and high concentration trains were statistically significant in most cases. As described in Chapter 1, this is consistent with theoretical expectation.
2. As with conventional treatment, removal of *Cryptosporidium* oocysts and *Giardia* cysts by the unit process of granular media filtration correlated with pathogen concentration at the filter influent. Lower pathogen removals were observed at lower pathogen concentrations.
3. When comparing previously published data with the results of this study, it is apparent that conventional treatment can achieve removals of 1 to 5 log for either *Cryptosporidium* oocysts or *Giardia* cysts. As expected, differences in raw water quality, treatment operation, and influent microorganism concentrations can all contribute to the reported differences in the pathogen removal capabilities of conventional treatment.
4. Continuous sampling allowed for the calculation of log removals in most cases when grab sampling produced non-detectable concentrations of *Cryptosporidium* oocysts or *Giardia* cysts. This approach provided an opportunity to evaluate pathogen spike doses of approximately 10^1 pathogens/L.
5. For raw water and filtered water samples, the results obtained by microscopy were not significantly different from those obtained by FC sorting-microscopy. However, settled water samples could not be sorted by the flow cytometer due to clogging of the flow cell. Therefore, FC sorting-microscopy appears to be a useful method for samples with relatively low particle content.
6. Mean log removal of *Giardia* cysts was significantly greater than mean log removal of *Cryptosporidium* oocysts through conventional treatment and through the filtration step. However, there was little difference between the two pathogens with respect to log removal through sedimentation.

RECOMMENDATIONS FOR WATER UTILITIES

The first of the conclusions described in the previous section has important implications for the drinking water community. The USEPA currently allows a 3-log credit for *Cryptosporidium* oocyst removal in conventional surface water treatment plants. For the water supply evaluated in this study, this level of removal was only observed at initial oocyst concentrations that were many orders of magnitude higher than the 10^{-3} to 10^{2} oocysts/L that are typically observed in a drinking water supply (LeChevallier *et al.* 1991, LeChevallier and Norton 1995, States *et al.* 1997, LeChevallier *et al.* 2003). Thus, it is plausible that some surface water systems may not achieve the level of *Cryptosporidium* oocyst removal they are given credit for.

These conclusions were based on experiments conducted with a water supply that has a relatively low particle content and a moderate dissolved organic carbon content. Similar experiments conducted on other water supplies may be expected to produce different conclusions. For example, sedimentation and filtration performance can both be expected to depend on total particle content and its influence on aggregation kinetics. Total particle content depends on raw water particle content, pathogen spike dose, and the amount of aluminum hydroxide particles created by alum addition. In this study, variations in pathogen spike dose produced significant variations in total particle content. For example, a pathogen spike of 10^{6} pathogens/L changed the influent total particle content from approximately 1.0 x 10^{6} particles/L to 2 x 10^{6} particles/L. Therefore, for the water evaluated in this study, the high pathogen spike of 10^{6} pathogens/L increased the influent particle concentration by a factor of 2. However, national average source water particle concentrations of 22,000/mL (2.2 x 10^{7} particles/L) have been reported in previous AwwaRF-funded research (McTigue *et al.* 1998). In pilot-scale investigations of waters like these, a pathogen spike of 10^{6} pathogens/L will only change the influent total particle concentration from 2.2 x 10^{7} to 2.3 x 10^{7} particles/L. Thus, the spiked water would have a total particle concentration that is only 1.05 times higher than the typically observed particle concentration. Because of the site-specific nature of particle content, water utilities are encouraged to participate in research studies that elucidate the impacts of overall particle content and particle characteristics on pathogen removal in conventional treatment processes. As always, utilities are strongly encouraged to maintain optimal operational conditions and to be vigilant in monitoring of filter effluent water quality.

The water industry is also encouraged to participate in risk assessments of the conclusions presented in this report. For example, studies with our pilot plant show an estimated 3-log removal of *Cryptosporidium* oocysts with a spike dose of $10^{5.5}$ oocysts/L, suggesting an average filter effluent concentration of $10^{2.5}$ oocysts/L. Meanwhile, an estimated 1-log removal is expected with a spike dose of 10^{1} oocysts/L, suggesting an average filter effluent concentration of 10^{0} oocysts/L. This latter filter effluent concentration is not as low as the 10^{-2} oocysts/L predicted by a 3-log removal credit. Therefore, a reassessment of public health risks is warranted, both now and after completion of studies recommended in the above paragraph.

Continuous separation channel centrifugation was shown to facilitate the understanding of pathogen removal at concentrations closer to the realistic levels that occur in natural water. Because the instrument allows for processing large volumes of filtered water, it can substantially reduce the uncertainty of calculating log removal due to low pathogen concentrations. Utilities are encouraged to work with manufacturers of these devices to assess the possibility of using them for monitoring of conventional treatment performance.

CHAPTER 1
INTRODUCTION

BACKGROUND

Several studies have evaluated the removal of *Cryptosporidium* oocysts, *Giardia* cysts, and other emerging waterborne pathogens in conventional drinking water treatment. For example, Harrington *et al.* (2001) observed 2.1 ± 0.2 log removal of *Cryptosporidium parvum* oocysts while Nieminski and Ongerth (1995) reported 3.0 ± 0.6 and 3.4 ± 0.7 log removal of *Cryptosporidium* oocyst and *Giardia* cysts, respectively. Based on these and other studies (e.g., Patania *et al.* 1995, McTigue *et al.* 1998, Emelko *et al.* 2000, Huck *et al.* 2001), the USEPA has established that conventional treatment achieves an average 3-log removal of *Cryptosporidium* oocysts (USEPA 2003). As a result, when the second stage of the Long-Term Enhanced Surface Water Treatment Rule was finalized in early 2006 (USEPA 2006), conventional treatment plants received a 3-log credit for oocyst removal if they met the filtration requirements of the rule. Spike concentrations used in these experiments are 6 to 8 orders of magnitude higher than those observed in practice. Occurrence surveys have shown that *Cryptosporidium* oocysts were detected in more than 63 percent of raw water locations sampled. For those cases where oocysts were detected, the geometric means were <0.001 to 3 oocysts/L, with a highest observed concentration of 484 oocysts/L (LeChevallier *et al.* 1991, LeChevallier and Norton 1995, States *et al.* 1997, LeChevallier *et al.* 2003). Thus, there is considerable uncertainty about pathogen removals that would be attained under more realistic situations. The scientific reasoning behind this uncertainty can be found by reviewing the theoretical analysis of particle concentration effects on particle separation kinetics.

Flocculation rate equations have been derived for several particle collision mechanisms by assuming that flocculation has second order kinetics in which the rate of collision between two differently sized particles is proportional to the product of the concentrations of the two differently sized particles (Smoluchowski 1917, Hunt 1982, Lawler *et al.* 1983, Han and Lawler 1992). For example, if two suspensions have an identical relative size distribution, but one has a concentration ten times the other, then the one with the higher concentration will flocculate 100 times more quickly. Hence, flocculation can be expected to be more effective at particle aggregation with higher particle concentrations.

Particle concentration also affects the rate of flocculant (Type 2) settling, the type encountered in conventional water treatment (Thomas *et al.* 1999, Smoluchowski 1917). Because flocculation rate increases with increasing particle concentration, the sedimentation rate will also increase when flocculant settling is encountered. In addition, with increased particle concentration, the free area between particles is reduced, causing greater interparticle fluid velocities and alteration of flow patterns around particles (Patwardhan and Chi 1985, Ekie-Jezewska *et al.* 1999, Burger *et al.* 2000). Dugan *et al.* (2001) evaluated *Cryptosporidium* oocyst removals in a pilot-scale conventional treatment plant with natural turbidity variation. The results of this study showed that higher raw water turbidity led to increased removal of *Cryptosporidium* oocysts through coagulation, flocculation, and sedimentation. Improved removals of *Aeromonas hydrophila* and *Escherichia coli* 0157:H7 were observed when turbidity was increased by spiking soil upstream of pilot-scale conventional treatment (Harrington *et al.*

2001). However, in the same experiment, no significant change in *Cryptosporidium* oocyst removal was observed.

Although particle concentration is known to affect particle retention in porous media (O'Melia and Ali 1978, Vigneswaran and Chang 1989), the originally proposed clean-bed filtration model suggested that the ratio of effluent concentration to influent concentration was independent of input concentration (see Equation 1, Iwasaki 1937). Yao *et al.* (1971) revealed that the ratio of effluent to influent concentration depended on the frequency of collisions between filter media (collectors) and particles moving through the filter. Thus, they proposed Equation 2, in which the value of λ is a function of a single-collector collision frequency (η) and an empirically derived collision efficiency (α). Depending on the physical and chemical characteristics of particles and collectors, η and α have been experimentally observed to increase with increasing particle concentration, which leads to an increase in rate of deposition. Prasanthi *et al.* (1997) found that $\alpha\eta$ increased with an increase in influent particle concentration. The variation of $\alpha\eta$ with influent concentration was also noticed during experiments with particles significantly larger than 1 µm (Chang 1989). Thus, the ratio of effluent to influent particle concentration can be expected to decrease with increasing influent particle concentration. In other words, log removal can be expected to increase with increasing influent particle concentration.

$$\frac{C}{C_0} = \exp[-\lambda L] \qquad (1)$$

Where

$C =$ colloid concentration exiting the packed bed.
$C_0 =$ colloid concentration entering the packed bed.
$\lambda =$ filtration coefficient.
$L =$ the thickness of the packed bed.

$$\lambda = \frac{3}{2}\frac{(1-\theta)}{d_c}\alpha\eta \qquad (2)$$

Where

$d_c =$ the diameter of collector (porous media grains).
$\theta =$ the porosity of the packed bed.
$\alpha =$ the collision efficiency.
$\eta =$ the collision frequency.

Several studies have evaluated the impacts of particle concentration when particles were seeded directly into the filters rather than being seeded in an upstream treatment process. When Prasanthi *et al.* (1997) used 0.46 µm latex microspheres to study the effect of particle concentration on the entire cycle of filtration, they found that log removal increased as microsphere concentration increased. With an initial concentration of 1.4×10^9 microspheres/L and 1.9 x 10^{11} microspheres/L, 0.3 log and 1.2 log removals were achieved, respectively. Direct filtration of inactivated *Cryptosporidium parvum* oocysts was studied by Emelko *et al.* (2001). They found median removals of 4.5 log and 4.8 log when spike doses of ~1 x 10^6 oocysts/L and ~5 x 10^4 oocysts/L were used, respectively. They concluded that seeding concentration had little impact on *Cryptosporidium parvum* oocyst removal by filtration.

OBJECTIVE

The overall goal of this study was to evaluate pathogen removals in conventional water treatment at pathogen concentrations nearer to those found in the aquatic environment. The specific objectives of this study were to:

1. Measure the removal efficiencies of *Cryptosporidium* oocysts and *Giardia* cysts that have been seeded at more realistic concentrations of 10^1 pathogens/L and describe removal efficiency as a function of seeded pathogen concentration over a range of influent concentrations from 10^1 to 10^5 pathogens/L.

2. Evaluate the differences between grab and composite sampling in quantifying pathogen removals.

3. Determine if particle sorting by flow cytometry is an alternative way of processing samples prior to microscopic enumeration of pathogens in natural water.

CHAPTER 2
METHODS

GENERAL APPROACH

This project evaluated pathogen removal with a unique combination of experimental and concentration methods. In order to characterize the removal of waterborne pathogens, experiments were conducted with different initial pathogen concentrations and sampling methods at the University of Wisconsin (UW) pilot plant. To minimize the hindrance of raw water quality as a confounding variable, treatment trains were run in parallel to facilitate better comparison of treatment performance between trains due to the effect of pathogen spike dose. Continuous separation channel centrifugation was used to concentrate pathogens while immunofluorescence microscopy was used for pathogen enumeration. Particle sorting with flow cytometry prior to microscopic enumeration (FC sorting-microscopy) was compared with the more conventional method of microscopic enumeration.

EXPERIMENTAL METHODS

Six experiments were conducted with one in November 2004, one in January 2005, one in March 2005, one in March 2006, one in December 2006 and one in March 2007. This section focuses on the experimental set-up for the January 2005 experiment, which is illustrated in Figure 2.1.

For all experiments, the pilot plant received raw water from Lake Mendota in Madison, Wisconsin. The raw water had a moderate to high dissolved organic carbon (DOC) concentration of 3 to 6 mg/L, a low specific UV absorbance (SUVA) of 1 to 2 L/mg·m, a high alkalinity of 250 to 330 mg/L as $CaCO_3$, a relatively low turbidity of 2 to 5 ntu, and total particle counts of 1,000/mL to 9,000/mL. These raw water particle counts are low when compared to the national average source water particle concentration of 22,000/mL (McTigue *et al.* 1998).

The pilot plant included two parallel treatment trains, each with a design capacity of 1 gallon per minute (gpm). In previous studies, the parallel treatment trains have been simultaneously spiked with duplicated initial pathogen concentrations to evaluate train-to-train variability in performance (Harrington *et al.* 2001). These studies demonstrated that the experimental error between the two trains was ± 0.2 to ± 0.3 log removal, depending on the pathogen being studied. This error was observed to be random, with no treatment train consistently achieving better pathogen removal than the other. Therefore, both trains can be considered to achieve the same level of performance when operated under identical conditions.

As shown in Figure 2.1, raw water went through coagulation, flocculation, and sedimentation on each treatment train. Raw water was pumped to a constant head reservoir and flowed by gravity through two rapid mix tanks, three flocculation chambers, and one sedimentation basin on each train. Each rapid mix tank operated with a 1 minute nominal detention time at design capacity and had an estimated velocity gradient of 750 sec^{-1}. Pathogen cocktail was fed in a step dose to the first rapid mix chamber and alum was fed to the second rapid mix chamber. At the 1 gpm flow rate, each flocculation chamber operated with a nominal detention time of 10 minutes for a total flocculation time of 30 minutes. Estimated velocity gradients were 70 sec^{-1}, 40 sec^{-1}, and 10 sec^{-1} for the first, second, and third flocculation

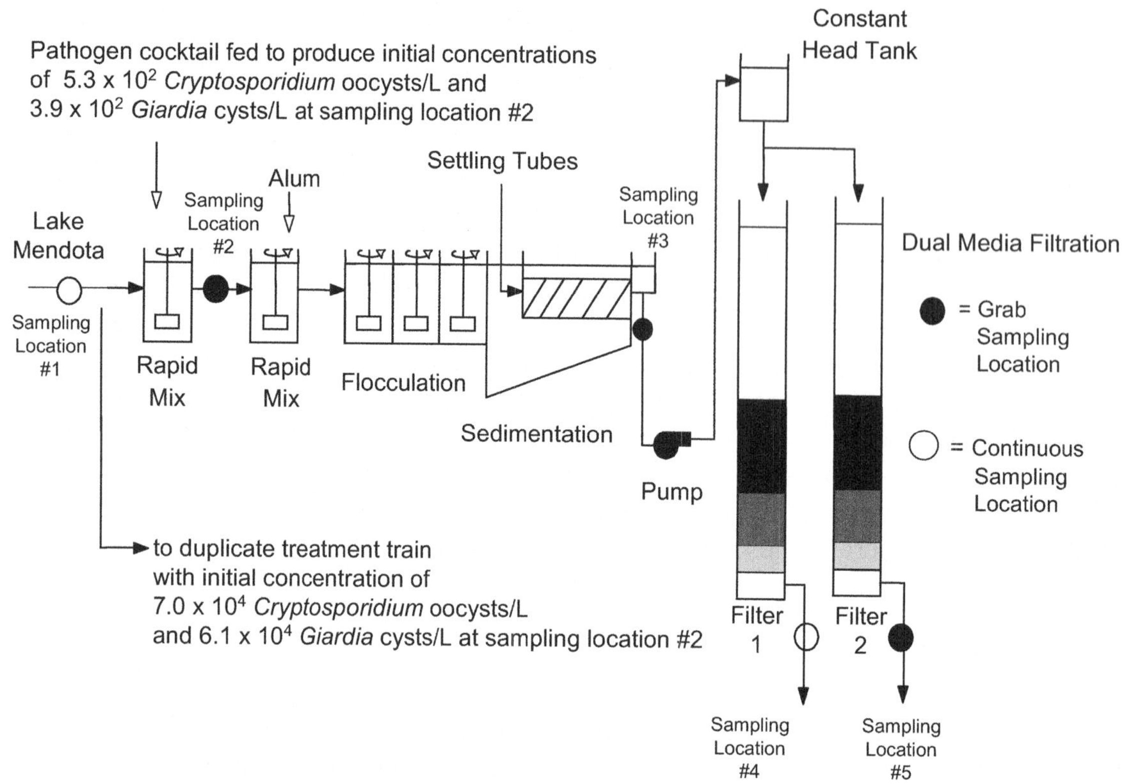

Figure 2.1 Process flow diagram for the January 2005 experiment

chambers, respectively. Water was distributed to each flocculation chamber and to the sedimentation basin through perforated walls. The sedimentation basin had a nominal detention time of approximately 2 hours at 1 gpm and included a tube settler module. The overflow rate of the sedimentation basin was estimated to be 70 gallons per day per square foot (gpd/ft^2). Pumping of settled water was provided because of space limitations in the laboratory. Settled water was distributed to a rack of two granular media filter columns constructed of clear polyvinylchloride with a diameter of 3 inches and a height of 10 feet. Dual media filtration with 18 inches of crushed anthracite over 12 inches of sand was used in all experiments. Characteristics of the filter media were described by Harrington *et al.* (2001). Wall effects were assumed to be minimal because the ratio of column diameter to filter medium diameter was greater than 70:1 for all filter media tested (Lang *et al.* 1993). For all experiments in this study, a constant filtration rate was maintained at 4 gpm/ft^2.

Previous experiments were conducted to evaluate hydraulic residence time distributions and to evaluate the influence of settled water pumping on turbidity removal (Harrington *et al.* 2001). The hydraulic residence time distribution showed that filter effluent samples could be collected no earlier than five hours after initiation of pathogen feed to the raw water supply. The other tests indicated that settled water pumping had no effect on turbidity removal, provided that the sedimentation basin was cleaned out one day prior to a challenge experiment.

In all experiments, the raw water was spiked with a pathogen "cocktail" that contained a mixture of *Cryptosporidium parvum* oocysts and *Giardia lamblia* cysts to facilitate direct comparisons between the two microbes. The cocktail was made by spiking the desired amount of pathogens into 15 L of Lake Mendota water in a high-density polyethylene (HDPE) container. This 15 L cocktail was continuously fed to each pilot train at a flow rate of 6.3 mL/min. A mixing system was designed to help maintain a uniform suspension of pathogens within the container, which allowed a uniform feed concentration of pathogens throughout the experiment. The mixing was provided by a magnetic stirring bar. Tables 2.1 and 2.2 summarize the spike doses used and raw water quality conditions encountered during the six experiments.

The spiked concentration was evaluated by collecting grab samples at Sampling Location 2 (see Figure 2.1) at three times during the experiment. These spiked raw water pathogen concentrations were reported as averages of counts obtained by both microscopy and FC sorting-microscopy (see page 13 for a description of these two enumeration methods). All spiked raw water pathogen concentrations shown in this report were average concentrations ± the standard deviation for three samples collected during the experiment (sample collection times are described later in this section). With two analytical methods and three sampling times, there were six values of raw water concentration for each treatment train. Therefore, the reported spike dose is the average and standard deviation of these six values.

For each train, the alum dose was selected as the dose that produced the lowest turbidity in jar tests where the source water was supplemented with test pathogens. This was done to ensure that the selected pilot-scale coagulant dose accounted for the concentration of the pathogen feed. Therefore, both treatment trains were run with optimal coagulation conditions.

SAMPLING PROTOCOL

As shown in Figure 2.1, samples were collected from four locations within each treatment train (see Sampling Locations 2 through 5). In addition to these 8 locations, samples were also collected from a location representing the raw water upstream of the pathogen spike (see Sampling Location 1). Therefore, a total of 9 locations were sampled for each experiment. As described later, these locations were sampled at multiple times during an experiment, producing more than 20 samples per experiment.

Figure 2.1 shows that this study investigated the use of two sampling methods (grab and continuous) for quantification of pathogens in the filter effluent. Grab sampling is useful for observing unusual conditions that may exist during a filter run by providing a snapshot of the filter effluent quality at a particular instant in time. However, the volume of water collected in a grab sample is limited to the 1 L to 10 L range, thereby limiting the number of pathogens in the collected sample. Also, due to the expense of pathogen analysis, only a few grab samples can be collected during a filter run and this does not necessarily provide an adequate description of the average performance throughout the filter run. Continuous sampling over the entire filter run does not allow for an assessment of variability in filter performance, but it offers the advantage of being able to collect a significantly larger volume of water (e.g., more than 100 L) and, hence, more pathogens for analysis. The resulting sample reflects the average characteristics during the filter run, which is an indication of overall performance for the treatment system.

The number of samples collected from each location is detailed in Figure 2.2. A continuous sample was collected from the raw water to evaluate the background concentration of pathogens in Lake Mendota (*Cryptosporidium* oocysts and *Giardia* cysts were not detected in this sample during any of the experiments). In each train, grab samples were collected immediately downstream of the location where the pathogens were spiked into the treatment

Table 2.1
Summary of pathogen concentrations

Experiment	Train 1		Train 2	
	Spiked *Cryptosporidium* Concentration$^{\gamma}$ oocysts/L	Spiked *Giardia* Concentration$^{\gamma}$ cysts/L	Spiked *Cryptosporidium* Concentration$^{\gamma}$ oocysts/L	Spiked *Giardia* Concentration$^{\gamma}$ cysts/L
November 2004	4.6×10^4 $\pm 1.4 \times 10^4$	7.0×10^4 $\pm 1.5 \times 10^4$	1.3×10^3 $\pm 1.3 \times 10^3$	6.8×10^2 $\pm 1.0 \times 10^2$
January 2005	5.3×10^2 $\pm 1.0 \times 10^2$	3.9×10^2 $\pm 0.5 \times 10^2$	7.0×10^4 $\pm 0.8 \times 10^4$	6.1×10^4 $\pm 1.2 \times 10^4$
March 2005	5.7×10^3 $\pm 1.4 \times 10^3$	4.4×10^3 $\pm 1.6 \times 10^3$	1.6×10^1 $\pm 1.7 \times 10^1$	2.2×10^1 $\pm 2.2 \times 10^1$
March 2006	2.3×10^1 $\pm 2.5 \times 10^1$	6.4×10^0 $\pm 3.2 \times 10^0$	3.0×10^3 $\pm 0.6 \times 10^3$	2.2×10^3 $\pm 0.7 \times 10^3$
December 2006	2.7×10^5 $\pm 1.2 \times 10^5$	3.2×10^5 $\pm 1.3 \times 10^5$	2.7×10^1 $\pm 1.3 \times 10^1$	2.4×10^1 $\pm 1.6 \times 10^1$
March 2007	1.7×10^1 $\pm 1.2 \times 10^1$	1.8×10^1 $\pm 1.5 \times 10^1$	2.3×10^5 $\pm 1.1 \times 10^5$	2.4×10^5 $\pm 1.4 \times 10^5$

γ Averaged concentration ± standard deviation

Table 2.2
Summary of raw water turbidities and alum doses

Experiment	Train 1		Train 2	
	Raw Water Turbidity$^{\zeta}$ ntu	Alum Dose mg/L	Raw Water Turbidity$^{\zeta}$ ntu	Alum Dose mg/L
November 2004	1.58 ± 0.05	80	1.63 ± 0.10	70
January 2005	1.59 ± 0.03	50	1.60 ± 0.06	60
March 2005	0.47 ± 0.03	60	0.47 ± 0.05	60
March 2006	0.61 ± 0.25	70	0.53 ± 0.06	80
December 2006	3.50 ± 1.02	80	3.60 ± 1.16	70
March 2007	1.19 ± 0.17	70	1.22 ± 0.13	80

ζ Measured downstream of pathogen spiking (averaged concentration ± standard deviation)

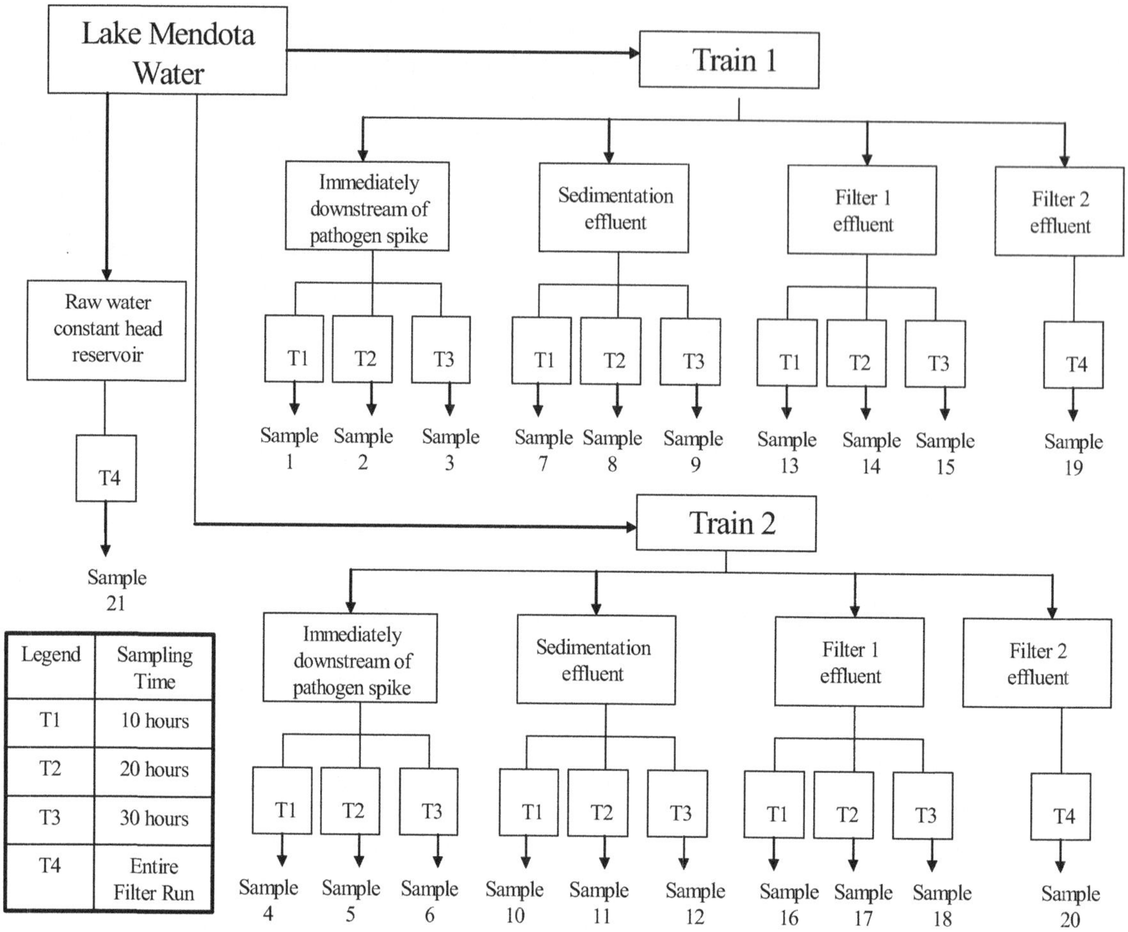

Figure 2.2 Schematic diagram of samples collected

plant (see sampling location #2 on Figure 2.1). This location was sampled at three separate times during an experiment (T1, T2, and T3 on Figure 2.2), which allowed verification of the actual spike dose and confirmation that the pathogen feed was operating at steady-state. The next sampling location was located at the effluent of the sedimentation basin (see sampling location #3 on Figure 2.1). Grab samples were also collected from this location at three times during an experiment (T1, T2, and T3 on Figure 2.2). This allowed the calculation of removals through sedimentation and confirmed that the treatment process upstream of filtration was achieving steady-state removals during the experiment. The remaining locations allowed quantification of removals through filtration and to compare grab sampling with continuous sampling. For the filter effluent, grab samples were also collected at three times during the experiment (T1, T2, and T3 on Figure 2.2). The continuous sampling method was run for the duration of the filter run, and produced a single composite sample from the entire filter run. As shown in Figure 2.2, there were a total of 21 samples for each experiment. Analytical error was quantified by duplicating the analysis of 10% of the samples.

Grab samples were collected at carefully selected times based on filtration theory and on experience with the typical day-to-day operation of the pilot plant. Ginn *et al.* (1992) have proposed a conceptual filtration model that consists of four phases: (1) filter ripening, (2) effective filtration, (3) breakthrough, and (4) wormhole flow. Filter ripening occurs when filter effluent turbidity decreases with time at the start of a filter run. Effective filtration is the second phase of operation, which begins when filter effluent turbidity is stabilized. With increasing particle deposition in the filter over time, filter effluent turbidity starts to breakthrough during the third phase. After breakthrough, preferential tubular passage (wormhole) of flow becomes significant. This fourth phase occurs when the capacity of the filter bed is almost exhausted. The effects of filter run time on pathogen removal have been studied in the UW pilot plant (Harrington *et al.* 2001). Results suggest that sampling must be carefully selected to appropriately assess pathogen removal during times when the filter effluent would be delivered to customers. The UW pilot plant typically achieves breakthrough after 40 hours of filter run time. Therefore, all samples were collected prior to this time, during the effective filtration phase.

In all experiments, feeding of the pathogen cocktail was initiated at a desired time. As noted earlier, tracer studies revealed that pathogen samples could not be collected from the filter effluent until 5 hours after pathogen spiking was initiated. Once this five-hour period was complete, each filter was backwashed and placed back into service. Grab samples were then collected at times that reflected the effective filtration phase. For example, the first sample from each grab sample location was collected 5 hours after the ripening phase was completed, based on results of turbidity monitoring during the experiment. The remaining grab samples were collected at 10 hour intervals after the first sample was collected. Continuous sampling began at the end of the ripening phase and was terminated at 35 hours of filter run time to avoid collecting samples from filters that had reached breakthrough.

The volume of the grab samples was 10 L and each grab sample was processed in accordance with the flow diagram shown in Figure 2.3. The first step of grab sample processing was concentration of the sample with continuous separation channel centrifugation (CSCC), which reduced the sample volume from 10 L to 200 mL. The operating protocol used for the CSCC is described later. After sample concentration, the sample was split into four aliquots of equal volume. One aliquot was sent to the Wisconsin State Laboratory of Hygiene (WSLH) for analysis with FC sorting-microscopy and another was sent to the Marshfield Clinic Research Foundation (MCRF) for analysis with microscopy. The remaining two aliquots were placed into archival storage at 4°C.

For the continuous samples, the centrifuges were connected to the filter effluent pipe and the centrifuges were continuously run over the entire phase of effective filtration. Continuous sampling was halted at 30 hours of sampling time (35 hours of filter run time) and a total sample volume of 288 L. The CSCC process reduced the 288 L sample volume to 200 mL, which was also split into four aliquots as shown in Figure 2.3. Based on the volume of water processed, the continuous sampling protocol had a detection limit that was one to two orders of magnitude lower than the detection limit offered by grab sampling. To verify the capacity of the centrifuge belt, samples were collected from the centrifuge effluent during an experiment. This effluent was centrifuged again and no pathogens were found in the samples, suggesting that the centrifuges had not reached breakthrough by the end of the 30 hour sampling time.

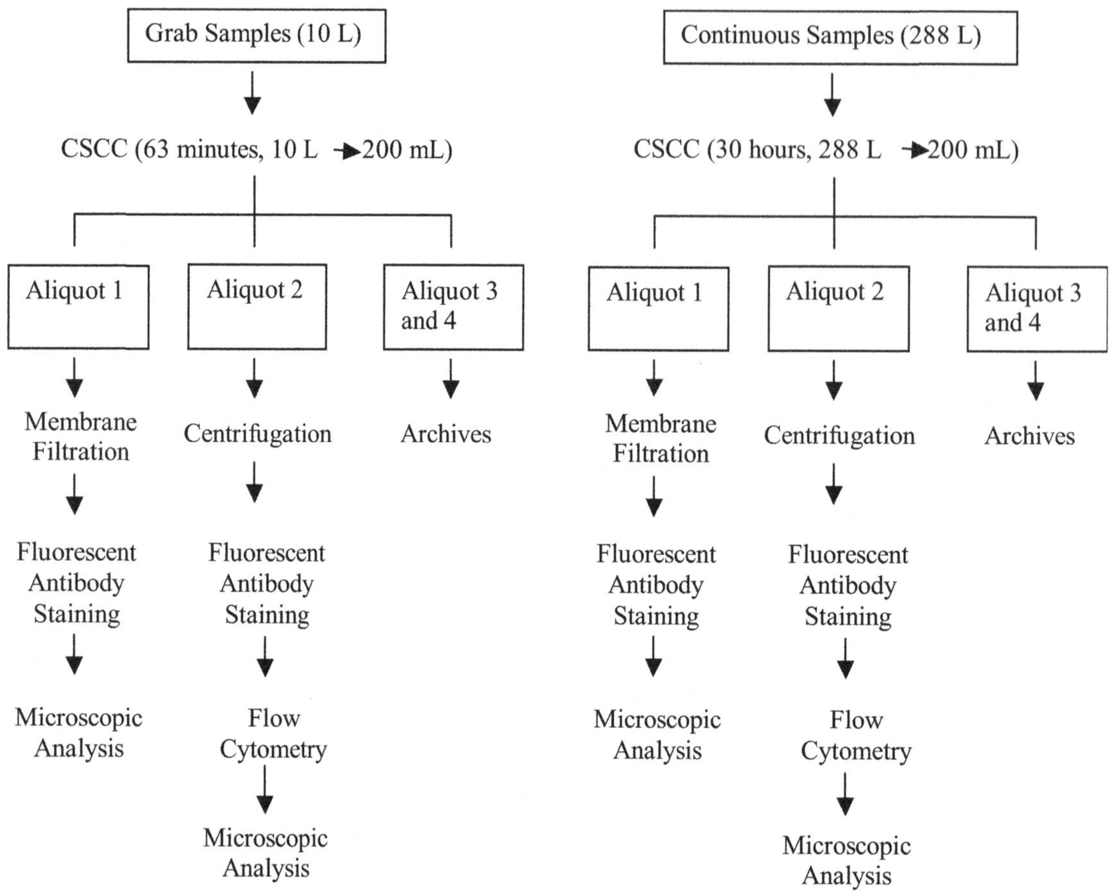

Figure 2.3 Schematic diagram of sample processing and analysis

CONCENTRATION OF PATHOGENS BY CONTINUOUS SEPARATION CHANNEL CENTRIFUGATION

Continuous separation channel centrifugation was performed according to the procedures described in Borchardt and Spencer (2002). MBR Water Separators (Molecular Biology Resources, Inc., Milwaukee, WI) were used to concentrate waterborne pathogens in all samples. All the separation channels were sterile, disposable and used only once. The separation channel design of the MBR is described by Borchardt and Spencer (2002).

By using a peristaltic pump, samples were fed into the separation channel through an input tube using one of the built-in centrifuge pumps at a feed rate of 160 mL/min. Rotor speed was set so that the relative centrifugal force at the channel width midpoint was approximately 900 x *g*. Pathogens were retained in the separation channel and the supernatant was pushed out of the channel through the output line. An inner wall separates the channel inlet from the outlet so that water pumped into one end must traverse the entire channel length before exiting. After centrifugation, the channel was removed from the rotor and the concentrate was drained into a sterile beaker. The separation channel was cut open and rinsed multiple times with 0.01% Tween 80 to dislodge pathogens that may have adhered to the inner chamber walls. Previous work using continuous separation channel centrifugation has shown that neither the surface epitopes (Hoffman *et al.* 2000) nor the viability (Borchardt and Spencer 2002) of *Cryptosporidium* oocysts or *Giardia* cysts were adversely affected during processing.

DETECTION OF WATERBORNE PATHOGENS BY IMMUNOFLUORESCENCE MICROSCOPY

Aliquots from each sample volume were filtered through a 1 μm absolute pore diameter, 25 mm, pre-blackened, polycarbonate membrane (Poretics, Livermore, CA) underlain with a polyester drain disc (Poretics), and held in a polycarbonate locking filter holder (Pall Gelman Laboratories, Ann Arbor, MA). The amount of sample filtered varied from 0.5 mL to 50 mL and was dependent on turbidity. Membranes were saturated with 70 μL of a 1:2 dilution of FITC-labeled monoclonal antibodies (Merifluor *Cryptosporidium/Giardia* Detection Kit, Meridian Bioscience, Cincinnati, OH) and transferred to a glass slide. For some samples, the entire filter was counted. For other samples, a portion of the filter was counted. The amount of the filter counted was dependent on the number of *Cryptosporidium* oocysts and on the number of *Giardia* cysts per field. Magnification was 200x, and presumed oocysts and cysts were confirmed at 400x based on external morphology. Counts of pathogens were performed with a Nikon Optiphot epifluorescent microscope (450-490 nm excitation, 505 nm dichroic mirror, 520 nm bandpass filter).

DETECTION OF WATERBORNE PATHOGENS BY FC SORTING-MICROSCOPY

Fifty mL aliquots were centrifuged at 3000 x *g* for 30 minutes in a Beckman J6M/E centrifuge using a JS5.2 swinging bucket rotor. The supernatant was aspirated and pellets (typically 150 to 500 μL volume) were transferred to 12 x 75 mm polypropylene tubes using multiple rinse steps to ensure that all of the sample was transferred. Samples were subjected to a second concentration by centrifugation at 1050 x *g* for 10 minutes and the supernatant was carefully aspirated. Sample volumes were precisely quantified and 1 to 200 μL of sample were blocked for 20 minutes to overnight with 90 μL PBS supplemented with 1% normal goat serum (Sigma, St. Louis, Missouri, Catalog Number G-9023). The amount of pellet volume assayed was dependent on the expected concentration of organisms (microscopic verification of greater than 1000 organisms will cause microscopy fatigue and introduce enumeration errors) and the amount of sediment in the sample (excess sediment will clog the instrument). This portion of the pellet volume was stained with FITC-conjugated monoclonal antibodies (CelLabs, Brookvale, Australia) against *Cryptosporidium parvum* and *Giardia lamblia*. Samples were processed with an Epics Elite flow cytometer (Beckman Coulter Corporation, Fullerton, CA) and events meeting predefined size and fluorescence criteria were sorted from nonfluorescent debris onto three-well glass microscope slides. Slides were allowed to air dry, cover-slipped with DABCO glycerol mounting media, and examined under 200X magnification using a Nikon E400 microscope equipped with fluorescence (460-500 nm excitation, 505 nm dichroic mirror, 510-560 nm bandpass filter). Cysts and oocysts were enumerated using fluorescence, size, and shape criteria. The entire sorted volume was examined by microscopy. Concentration factors for the oocyst and cyst pellets are shown in Appendix H.

RECOVERY CONTROLS

Recovery controls and matrix background checks were run at the onset of each experiment. For recovery controls, standards containing 500 oocysts and 500 cysts were seeded into 10 L grab samples of unseeded matrix water. The standards were prepared by personnel at the Wisconsin State Laboratory of Hygiene by using an Epics Elite flow cytometer (Beckman Coulter Corporation, Fullerton, CA). Three different matrix waters were tested: raw water, settled water and filtered water. These waters were collected at the onset of each experiment, before spiking pathogens into both treatment trains. After adding the standards to the matrix water, the samples were then processed using the procedures described earlier. The number of counted oocysts or cysts was then divided by the seed number to calculate the recovery efficiency (see pages 15 and 16 for count calculations). Results are shown in Appendix G.

All recovery control tests were done at concentrations of 50 oocysts/L and 50 cysts/L. Samples collected in the experiments had *Cryptosporidium* concentrations that ranged from 10^{-1} to 10^{5} oocysts/L. For *Giardia*, samples collected in the experiments had concentrations that ranged from 10^{0} to 10^{5} cysts/L. Previous research has shown that recovery efficiencies can depend on pathogen concentrations (Borchardt and Spencer 2002). Therefore, caution must be exercised when extrapolating recovery efficiencies from concentrations of 5×10^{1}/L in recovery controls to the wide range of values observed in the experimental samples. In Chapter 3, sensitivity analyses show that log removal comparisons between two treatment trains are relatively insensitive to recovery values.

SOURCES OF *CRYPTOSPORIDIUM* OOCYSTS AND *GIARDIA* CYSTS

Cryptosporidium parvum oocysts were obtained and purified using a cesium chloride gradient method from infected calf feces (Department of Veterinary Medicine, University of Arizona) while *Giardia lamblia* cysts were obtained from a commercial laboratory (Waterborne, Inc., New Orleans, LA). Parasites were assessed for overall quality by examination under differential interference contrast microscopy prior to use in any experiments. *Cryptosporidium* oocysts were used no longer than three months following the shed date. *Giardia* cysts were formalized (5% formalin) and used within a 3 month time period. The preferred approach in treatment studies is to use live pathogens because chemical treatments like formalin can alter the chemistry of the microbe surface. Unfortunately, live *Giardia* cysts were not available in sufficient quantities for these experiments.

TURBIDITY ANALYSIS

Effluent turbidity of one filter on each treatment train was measured by an on-line low range turbidimeter (Model 1720E, Hach Company, Loveland, Colorado). The turbidimeters were calibrated with prepared Stablcal™, stabilized formazin solution.

In addition to on-line turbidity monitoring of the two filter effluents, grab samples were collected for turbidity analysis from the 9 locations described earlier (see Figure 2.1). The turbidity of these samples was measured with a bench-top turbidimeter (Model 2100N, Hach Company, Loveland, Colorado) in accordance with Standard Method 2130 B (APHA *et al.* 1999). The turbidimeter was calibrated before every experiment with standard formazin suspensions (hydrazine sulfate and hexamethylene tetramine, Sigma, St. Louis, Missouri).

PARTICLE COUNT ANALYSIS

Filter effluent particle counts were measured for one filter on each treatment train by an on-line particle counter (Model 2200PCX, Hach Company, Loveland, Colorado). The particle counters were calibrated at the factory with NIST traceable polystyrene latex spheres (Duke Scientific, Fremont, California).

pH ANALYSIS

pH was measured with a plastic-bulb liquid-junction electrode (FB-10, Fisher Scientific, Pittsburgh, PA). The pH meter was calibrated before every experiment with pH 4, pH 7, and pH 10 buffers.

ANALYSIS OF UV254 ABSORBANCE

Ultraviolet absorbance was measured with a spectrophotometer (DR-4000V, Hach Company, Loveland, Colorado) in accordance with Standard Method 5910B (APHA *et al.* 1999).

DATA ANALYSIS

Calculation of Pathogen Concentrations

For samples analyzed by microscopy, the pathogen concentration of the collected sample was calculated using Equation 3.

$$N = \left(\frac{n_P f_F}{n_F V_F}\right)\left(\frac{n_A V_A}{V_S}\right) \tag{3}$$

Where:

- N = concentration of pathogen in the collected sample (pathogens/L).
- f_F = number of fields per filter. For the filters and microscopy conditions used in this work, $f_F = 816.3$ fields per filter.
- n_A = number of aliquots that the CSCC-concentrated sample was divided into. The 200 mL CSCC-concentrated sample was divided into four 50 mL aliquots for all experiments except for November 2004, in which 150 mL of CSCC-concentrated sample was divided into three 50 mL aliquots. Thus, $n_A = 4$ for all experiments except November 2004. For the November 2004 experiment, $n_A = 3$.
- n_F = number of fields on the filter that were counted by the microscopist.
- n_P = number of pathogens counted by the microscopist.
- V_A = volume of CSCC-concentrated aliquot that was analyzed. $V_A = 50$ mL for all experiments.
- V_F = volume of CSCC-concentrated sample aliquot that was filtered for counting. If the entire aliquot was filtered, then $V_F = 50$ mL.
- V_S = volume of sample collected. For grab samples, $V_S = 10$ L. For continuous samples, $V_S = 288$ L.

In Equation 3, the first parenthetical term is the concentration of pathogens in the CSCC-concentrated sample and the second parenthetical term is the inverse of the concentration factor achieved by the CSCC. The concentration factors achieved by the CSCC were 50 for the grab samples and 1440 for the continuous samples.

For samples analyzed by FC sorting-microscopy, the pathogen concentration of the collected sample was calculated using Equation 4.

$$N = \left(\frac{n_P}{f_A V_L}\right)\left(\frac{V_P}{V_A}\right)\left(\frac{n_A V_A}{V_S}\right) = \frac{n_P n_A V_P}{f_A V_L V_S} \tag{4}$$

Where:

$N =$ concentration of pathogen in the collected sample (pathogens/L).
$f_A =$ fraction of labeled sample that was analyzed. If all of the labeled sample was analyzed, then $f_A = 1$.
$n_A =$ number of aliquots that the CSCC-concentrated sample was divided into. The 200 mL CSCC-concentrated sample was divided into four 50 mL aliquots for all experiments except for November 2004, in which 150 mL of CSCC-concentrated sample was divided into three 50 mL aliquots. Thus, $n_A = 4$ for all experiments except November 2004. For the November 2004 experiment, $n_A = 3$.
$n_P =$ number of pathogens counted by the microscopist.
$V_A =$ volume of CSCC-concentrated aliquot that was analyzed. $V_A = 50$ mL for all experiments.
$V_L =$ volume of pellet that was labeled with FITC-conjugated antibody.
$V_P =$ volume of pellet achieved by swinging-bucket centrifuge.
$V_S =$ volume of sample collected. For grab samples, $V_S = 10$ L. For continuous samples, $V_S = 288$ L.

In Equation 4, the first parenthetical term is the concentration of pathogens in the pellet that was analyzed. The second parenthetical term is the inverse of the concentration factor achieved by the swinging-bucket centrifuge and the third parenthetical term is the inverse of the concentration factor achieved by the CSCC. As noted earlier, the concentration factors achieved by the swinging-bucket centrifuge are shown in Appendix H. The concentration factors achieved by the CSCC were 50 for the grab samples and 1440 for the continuous samples.

Calculation of Log Pathogen Removal

The log pathogen removals in this report were calculated using Equation 5.

$$Log\ pathogen\ removal = -Log_{10}\left(\frac{N}{N_0}\right) \tag{5}$$

Where

$N =$ concentration of pathogen in treated water.

N_0 = concentration of pathogen in source water.

Recovery efficiency was not included in the calculation of log removal for two reasons. First, as noted earlier, recovery efficiency was always evaluated at concentrations of 5 x 10^1 oocysts/L and 5 x 10^1 cysts/L. Previous research has shown that recovery efficiencies can depend on pathogen concentrations (Borchardt and Spencer 2002). No experiments were performed to evaluate recovery efficiency at the other pathogen concentrations observed during this study. Second, a comparison of log removals obtained without employing the recovery factor was made with log removals that were adjusted by the recovery factor (see the results chapter). This comparison revealed that the differences between the alternative calculations of log removal were less than the experimental error of 0.3 log removal for the equipment used in the study.

For all log removal calculations, the value of N_0 was an average of all sample concentration measurements obtained. In the case of log removals calculated for the entire treatment train, the value of N_0 was the average of six spiked raw water measurements obtained from three sample times and two analytical methods. The same was true for log removals calculated from the point of alum addition to the effluent of the sedimentation basin. As shown in the results chapter, the FC sorting-microscopy method was unable to measure concentrations for sedimentation effluent samples. Thus, log removals calculated for the unit process of granular media filtration were based on an N_0 that was equal to the average of three sedimentation effluent measurements (three sample times and one analytical method).

For those cases where a filter effluent sample had a pathogen concentration below the detection limit, the log removal was calculated by using the detection limit for the value of N. In some tables presented in Appendices C through F, these log removals were presented as "greater than" values. However, these values were never used in the *t*-tests or regression analyses described in the next section.

Statistical Analyses

Paired t-test

Paired, two-sample, two-tailed *t*-tests with a 95% confidence level were used to determine if there were statistically significant differences in mean log removal between:

- grab sampling and continuous sampling,
- *Cryptosporidium* and *Giardia*, and
- microscopy and FC sorting-microscopy.

The structure of the matched pair comparison statistical test is shown in Table 2.3. Difference pairs are independent, and the D_i values are independent of one another. Because $D_i = X_i - Y_i$ (for $i = 1, 2, \ldots, n$) constitutes a random sample from a population with mean difference = δ and variance = δ_D^2, hypotheses about δ can be tested using a one-sample *t*-test (Devore 2000). The following summary statistics can be used to perform hypothesis tests for the mean difference δ:

Table 2.3
Structure of matched pair comparison statistical test

Pair	Group 1	Group 2	Difference
1	X_1	Y_1	$D_1 = X_1 - Y_1$
2	X_2	Y_2	$D_2 = X_2 - Y_2$
.	.	.	.
.	.	.	.
.	.	.	.
n	X_n	Y_n	$D_n = X_n - Y_n$

$$\overline{D} = \frac{1}{n}\sum_{i=1}^{n} D_i \qquad (6)$$

$$s_D = \sqrt{\frac{\sum_{i=1}^{n}(D_i - \overline{D})^2}{n-1}} \qquad (7)$$

$$t = \frac{\overline{D} - \delta_0}{s_D/\sqrt{n}}, \text{ d.f.} = n - 1 \qquad (8)$$

Where

$\overline{D}$ = average difference between n pairs of experimental units.
s_D = standard deviation of the differences between n pairs of experimental units.
t = test statistic.
d.f. = degrees of freedom.
δ_0 = hypothesized mean difference (equal to zero for this report).

The summary statistics shown in Equations 6 through 8 can be used to test the hypothesis that $\delta \neq \delta_0$ with the rejection criteria for the null hypothesis given by $|t| \geq t_{\alpha/2}$. The parameter $t_{\alpha/2}$ is the upper $\alpha/2$ point of the t distribution with $\alpha = 0.05$ and $n - 1$ degrees of freedom. The rejection criterion is also determined by two times the area to the right of $|t|$ under the t curve with $n - 1$ degrees of freedom, which produces a statistic known as the p-value (two-tailed test). If the p-value < 0.05, the null hypothesis can be rejected at a significance level of 0.05 (i.e., the 95th percentile). T-tests for this research were performed using the data analysis function of Microsoft Excel.

Regression analysis

Simple regression. In this research, simple regression was used to assess the relationships between:

- results analyzed by FC sorting-microscopy and results analyzed by microscopy,
- removals of pathogens and filter run time, and
- removals of pathogens and initial pathogen concentration.

The simple linear model was used in this research:

$$y = \beta_0 + \beta_1 x \tag{9}$$

Hypothesis-testing was used to determine whether there was a statistically significant relationship between the tested parameters (x and y). One test used in the analysis was to determine if the slope (β_1) was significantly different from zero. For this, the model test was the test of H_0: $\beta_1 = 0$ versus H_a: $\beta_1 \neq 0$, in which case the test statistic value was the t-ratio $t = \hat{\beta}_1 / s_{\hat{\beta}_1}$. The rejection criterion was determined by two times the area to the right of $|t|$ under the t-curve with n - 2 degrees of freedom, which is called the p-value for the two-tailed test. If the p-value < 0.05, the null hypothesis can be rejected at a significance level of 0.05. In other words, if the p-value < 0.05, then the slope of the fitted line was considered to be nonzero (Devore 2000). A similar approach was used to test the significance of the intercept estimate.

Additional hypothesis-testing was used to determine whether the results analyzed by FC sorting-microscopy were the same as the results analyzed by microscopy. In other words, we tested the hypothesis that the slope of the regression line was equal to a value of unity. The model test was the test of H_0: $\beta_1 = 1$ versus H_a: $\beta_1 \neq 1$, in which case the test statistic value was the t-ratio $t = (\hat{\beta}_1 - 1)/s_{\hat{\beta}_1}$. The rejection criterion was determined the same way as the hypothesis-testing mentioned above. As with the paired t-test, the data analysis function of Microsoft Excel was used to conduct the test.

Multiple regression. In this research, multiple regression was used to assess the relationships between log removal and system variables such as initial pathogen concentration, raw water turbidity, alum dose, raw water temperature, sampling methods, treatment train, and raw water UV absorbance. First, simple regression was conducted for individual predictors, and parameters that resulted in zero R^2 were not included in the multivariable model. The general model used in this research was:

$$y = \beta_0 + \beta_1 x_1 + \beta_2 x_2 + \ldots + \beta_k x_k \tag{10}$$

Hypothesis-testing was used to determine whether there was a relationship between the tested parameters (x and y). The test used in the analysis was to determine if coefficient values (β_1, β_2, …, β_k) were significantly different from zero. For this, the model test was the test of H_0: $\beta_1 = \beta_2 = \ldots = \beta_k = 0$ versus H_a: at least one $\beta_i \neq 0$ (for $i = 1, \ldots, k$), in which case the test statistic value was the t-ratio $t = \hat{\beta}_1 / s_{\hat{\beta}_1}$. The rejection criterion was determined by two times the area to the right of $|t|$ under the t-curve with n – 2 degrees of freedom, which is called the p-value for the two-tailed test. If the p-value < 0.05, the null hypothesis can be rejected at a significance level of 0.05. In other words, if the p-value < 0.05, then the slope of the fitted line was considered to be nonzero. A similar approach was used to test the significance of the intercept estimate.

When the null hypothesis was not rejected, that meant there was a useful linear relationship between y and at least one of the predictors in the model. This did not mean that all predictors were useful, so additional steps were needed for model selection. In this research, stepwise regression was used with the backward elimination method at an F value $= 4$ as a criterion to eliminate non-significant parameters. For example, the parameter that has an absolute F value less than 4 would be eliminated while the other parameters would be retained. In this way, the algorithm continued until all absolute F values were at least 4 or greater. The model used was the one containing all predictors that were not eliminated. The F value of 4 was used because most F values at a 95 percent confidence level are near 4 (Draper and Smith 1998). Minitab 15 was used to conduct the test.

CHAPTER 3
RESULTS AND DISCUSSION

COMPARISON OF MICROSCOPY WITH FC SORTING-MICROSCOPY

As noted in the previous chapter, samples were analyzed with two different approaches. In one method, samples were filtered and the particles on the filter were then stained and counted manually under a microscope. In the other method, samples were concentrated by centrifugation. The concentrated particles were then stained and sorted by flow cytometry onto a microscope slide. Sorted oocysts and cysts were then counted manually under a microscope. For convenience, this report will refer to the first method as microscopy and the second method as FC sorting-microscopy.

Data from all six experiments were combined to compare the log counts of pathogens by FC sorting-microscopy and microscopy. These six experiments included a total of 132 samples, 19 matrix spikes, and 6 background samples, yielding a possible total of 157 matched pairs of FC sorting-microscopy and microscopy results. Of these 157 pairs, flow cytometry was not capable of processing 26 settled water samples and 2 filtered water samples due to plugging of the flow cytometer cell. For 17 matched pairs, both FC sorting-microscopy and microscopy were unable to detect either cysts or oocysts (all 6 of the background samples were in this group). In 9 more matched pairs, FC sorting-microscopy was able to detect pathogens when microscopy was not able to. In addition, there were 7 matched pairs for oocysts and 8 matched pairs for cysts in which microscopy was able to detect oocysts when FC sorting-microscopy was not able to.

The remaining data consisted of 96 and 95 paired observations for *Cryptosporidium* and *Giardia*, respectively. The paired, two-sample, two-tailed *t*-test was used to compare the two methods. The paired *t*-test results indicated that there was no significant difference between FC sorting-microscopy and microscopy for mean log oocyst counts (*p*-value = 0.60) or for mean log cyst counts (*p*-value = 0.08).

The 96 and 95 matched pairs are presented in Figures 3.1 and 3.2 for oocysts and cysts, respectively. These figures show paired results for all six experiments and include the matrix spikes described in Appendix G. For *Cryptosporidium* oocysts, the log counts by FC sorting-microscopy were correlated with the log counts from microscopy with a slope of 0.89. This slope had a standard error of 0.05 and a *p*-value less than 0.01 which suggests that the observed trend was significantly different from zero with more than a 95 percent level of confidence. The intercept of 0.26 had a p-value of 0.09, suggesting that the intercept was not significantly different from zero at a 95 percent confidence level.

A similar analysis of *Giardia* cyst counts produced a slope of 0.90 and an intercept of 0.16 (see Figure 3.2). When testing was done to evaluate whether the intercept and slope were different from a value of zero, the intercept and slope had *p*-values of 0.16 and less than 0.01, respectively, suggesting that the slope was significantly different from zero while the intercept was not.

Because the intercept was not significantly different from zero in either case, additional testing was done without an intercept to assess whether the slope of the observed trend was significantly different from a value of one. The results are shown in Figures 3.3 and 3.4 for *Cryptosporidium* oocysts and *Giardia* cysts, respectively. The slope shown in Figure 3.3 had a value of 0.97. When testing was done to determine if this slope was significantly different from

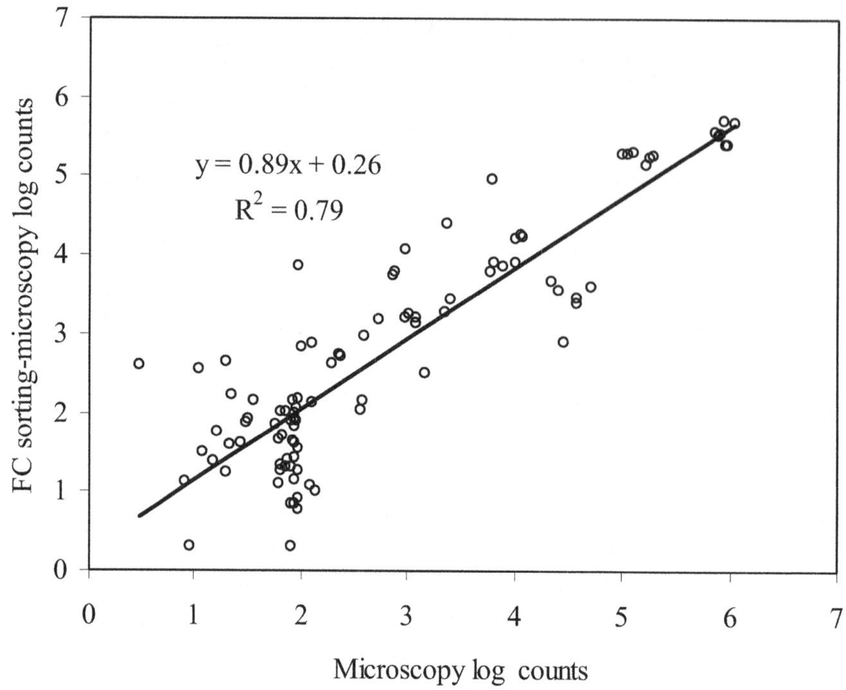

Figure 3.1 Comparison of log *Cryptosporidium* oocyst counts by FC sorting-microscopy with log *Cryptosporidium* oocyst counts by microscopy

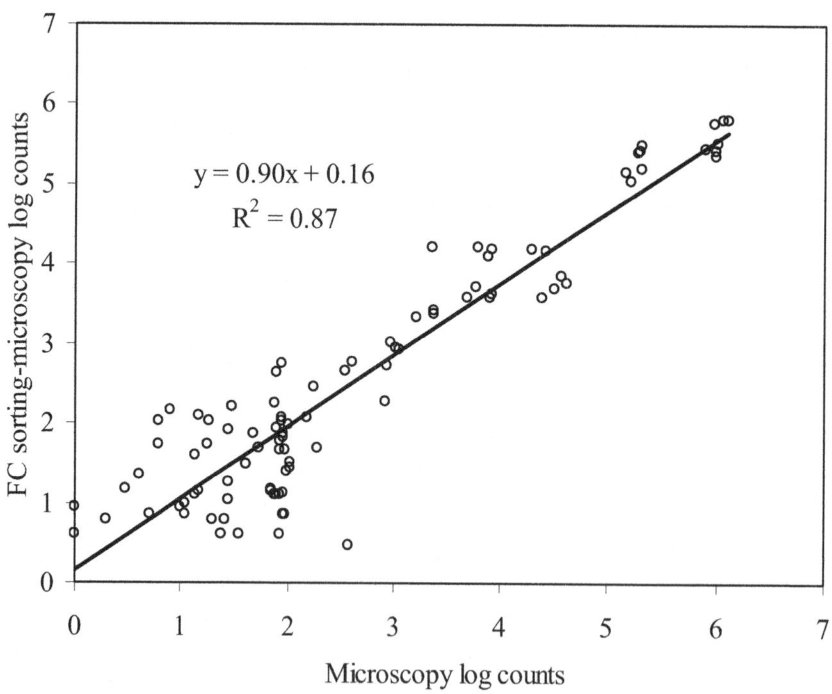

Figure 3.2 Comparison of log *Giardia* cyst counts by FC sorting-microscopy with log *Giardia* cyst counts by microscopy

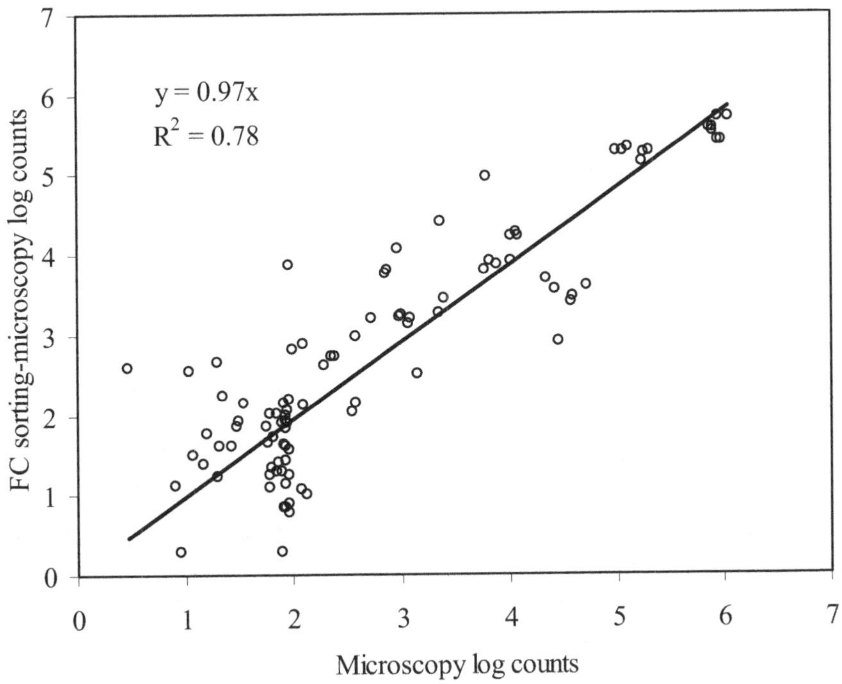

Figure 3.3 Comparison of log *Cryptosporidium* oocyst counts by FC sorting-microscopy with log *Cryptosporidium* oocyst counts by microscopy based on simple regression without intercept

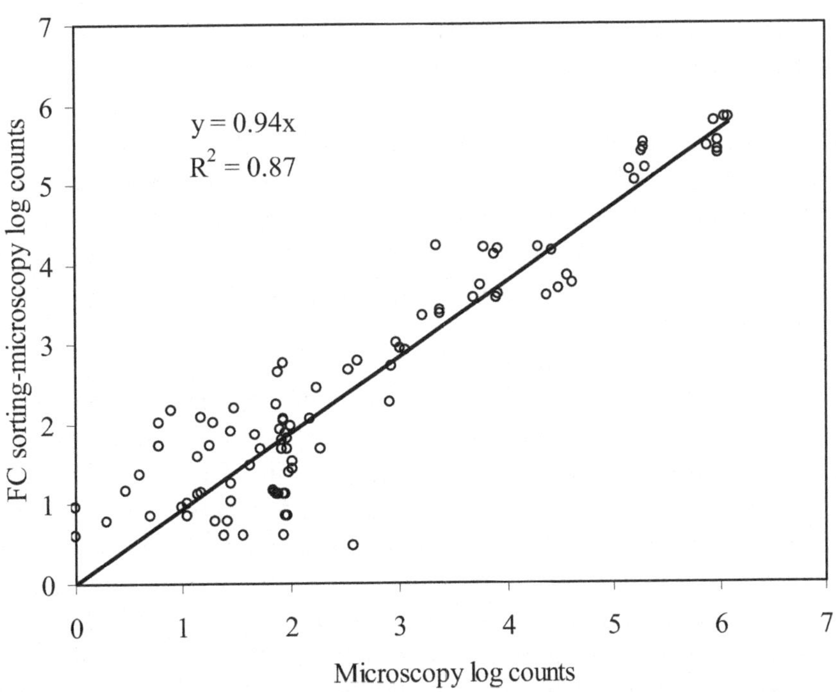

Figure 3.4 Comparison of log *Giardia* cyst counts by FC sorting-microscopy with log *Giardia* cyst counts by microscopy based on simple regression without intercept

a value of one, the p-value of the test was 0.08. This suggested that the slope was not significantly different from one at a 95 percent confidence level, confirming that the results from the two methods were not significantly different.

For *Giardia* cysts, the slope was 0.94 and testing suggested that this slope was significantly different from one at a 95 percent confidence level (p-value < 0.01). An assessment of the 95 percent confidence interval for the slope showed that the slope could be somewhere between values of 0.91 and 0.98. Although the 95 percent confidence interval is close to including the value of one, this test suggested that the analytical methods produced statistically different results. This is contrary to the results suggested by the paired t-test. While the statistical significance test showed that the slope was significantly different from a value of one, the practical significance of the difference can also be considered. Based on the slope of 0.94 shown in Figure 3.4, a log cyst count of 6.0 obtained from microscopy is expected to be associated with a log cyst count of 0.94*6.0 = 5.6 obtained from FC sorting-microscopy. This difference of 0.4 log is very close to the experimental error of the pilot plant.

Given the results described in this section, we conclude that the microscopy method was more appropriate than the FC sorting-microscopy method for samples collected from the effluent of the sedimentation basin. Therefore, for samples collected from the effluent of the sedimentation basin, only results from microscopy were used. However, for samples with lower particle content (i.e., raw and filtered waters), both methods gave a reasonably similar estimate of the pathogen concentration in a given sample. Therefore, counts from both methods were averaged for both raw and filtered water samples and these average counts were used for data analysis. Log removals calculated in this manner account for interlaboratory variability in measurement methods.

PATHOGEN REMOVAL FROM ALUM ADDITION THROUGH SEDIMENTATION

As shown in Appendices A and B, the November 2004 and January 2005 experiments suggested that higher log removals were achieved through sedimentation in treatment trains with higher initial pathogen concentrations. For the four remaining experiments, grab samples collected from the sedimentation basin effluent were below the detection limit in the treatment train with the lower initial pathogen concentration (see Appendices C through F). For these same experiments, log removals ranged from 0.8 to 1.6 in the treatment train with the higher initial pathogen concentration. However, because log removals could not be calculated for the low concentration trains, a direct comparison of performance between the two trains was not feasible in the latter four experiments.

To determine if general trends could be observed by combining results from all six experiments, results from the sedimentation basin effluent were used to generate Figures 3.5 and 3.6. As noted in Chapter 2, these figures do not include data from treatment trains that produced pathogen concentrations that were below the detection limit. Results were also compared with results obtained by other investigators. Figure 3.5 shows the results obtained for *Cryptosporidium* oocyst removals. This graph shows a trend of increasing log removal with increasing spike dose of *Cryptosporidium* oocysts. The slope of 0.12 shown in Figure 3.5 had a standard error of 0.03 and a p-value less than 0.01, which suggests that the observed trend was statistically significant at a 95 percent level of confidence. The R^2 value suggests that the initial oocyst concentration could explain 52 percent of the variance in the log removal data. This result also suggests that other factors played a role in determining the log oocyst removal

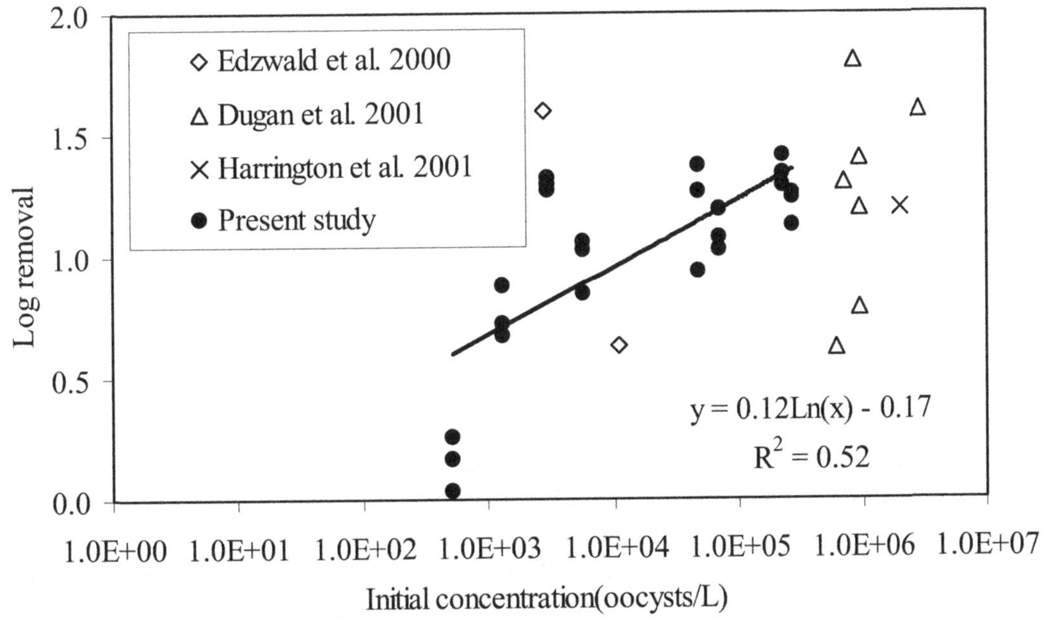

Figure 3.5 ***Cryptosporidium*** **oocyst removals through sedimentation versus initial concentration.**

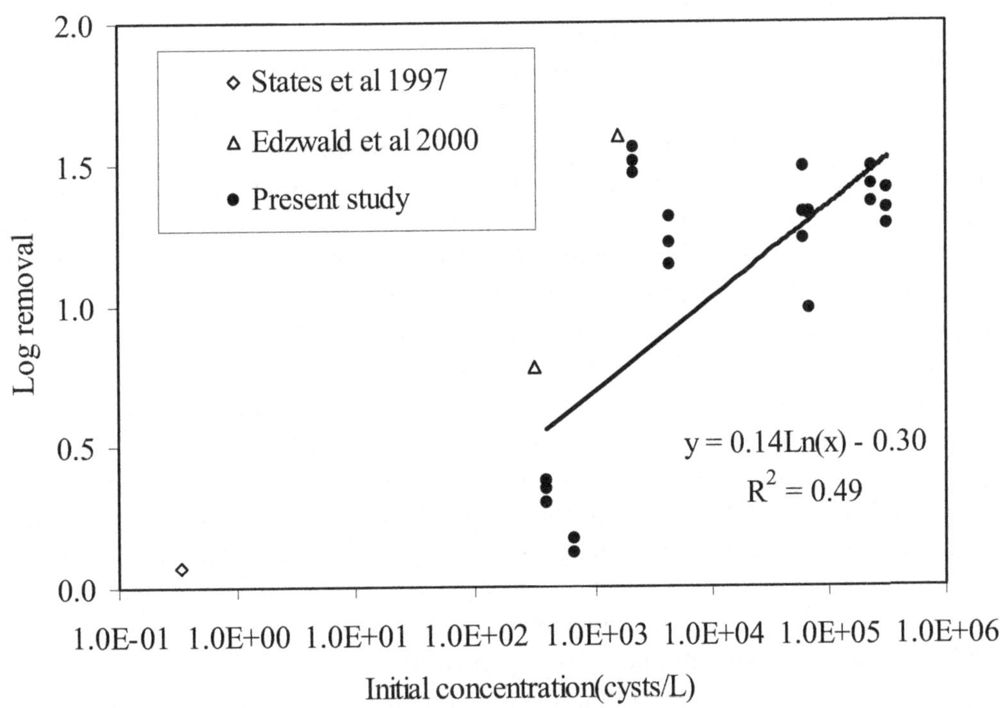

Figure 3.6 ***Giardia*** **cyst removals through sedimentation versus initial concentration.**

achieved through sedimentation. For example, other possible explanations for the variance in log removal could be differences in raw water quality or other experimental conditions from one experiment to another.

To test this, stepwise multiple regression was performed to determine if any of the raw water quality variables or other experimental conditions had a significant influence on oocyst removal. The result suggested that log removal of *Cryptosporidium* oocysts after sedimentation was positively correlated with both initial oocyst concentration and alum dose with a p-value less than 0.01. Because the p-values for the coefficient of the predictor variables were less than 0.01, the relationships were statistically significant. The model that contained both initial pathogen concentration and alum dose explained 73 percent of the variation in the data. This suggests that some of the variation shown in Figure 3.5 could come from the different alum doses used in the six different experiments. As explained earlier, jar tests were used to determine an optimal alum dose for each treatment train. Optimal alum dose depends on raw water quality parameters such as particle content and dissolved organic carbon content. The stepwise regression analysis did not suggest raw water turbidity or raw water UV254 as significant contributors to log oocyst removal through sedimentation. Increasing alum dose can have a significant influence on the surface chemistry of particles present in the source water, but it also increases the amount of aluminum hydroxide particles present in flocculation and sedimentation. The increased number of particles can, in turn, influence the rate of particle-particle interactions and the rate of particle aggregation.

As noted earlier, data from other studies are also shown on Figure 3.5. Quantitative analysis of these comparisons is often difficult because differences in raw water quality and treatment plant operations are often not reported. However, qualitative comparisons can be made in an effort to elucidate patterns that might require further study. For example, data from a previous pilot study with the same treatment plant were consistent with data from the present study (Harrington *et al.* 2001). Also, data from a full-scale operation showed that relatively low log-removals were observed through sedimentation when naturally-present oocyst concentrations were evaluated for the water supply in Pittsburgh, PA (States *et al.* 1997). Data from the present study lie within the range observed by Edzwald *et al.* (2000), who used a Connecticut water supply having turbidities of 0.5 to 1.5 ntu, TOC concentrations of 2.5 to 3.0 mg/L, and alum doses of 18 to 28 mg/L. They also used a pulse-dose spiking approach rather than a step-dose spiking approach like the one used in the present study. Using the Ohio River as the source, Dugan *et al.* (2001) observed a 1.2-log range of removals over a relatively narrow range of initial oocyst concentrations. Examination of the data collected by Dugan *et al.* reveals that log removal increased with increasing raw water turbidity over a raw water turbidity range of 1 to 100 ntu. The raw water particle count was 3.4 x 10^7/L to 5.3 x 10^9/L, raw water TOC concentration was 1.7 to 3.0 mg/L, and alum dose was 10 to 50 mg/L.

These observations, combined with the observations of the present study, suggest that overall particle load may have a significant influence on pathogen removal through sedimentation. The raw water turbidity in the present study only ranged from 2 to 5 ntu and raw water particle counts ranged from 1 x 10^6 to 9 x 10^6 particles/L upstream of the pathogen spike location. Meanwhile, the pathogen spike represented 10^3 to 10^6 particles/L when accounting for both oocysts and cysts. Therefore, the pathogens had a significant influence on the overall particle load entering the treatment plant in the present study.

Figure 3.6 shows the results obtained for *Giardia* cyst removal compared with results from previous studies. Like the oocyst removal data, the *Giardia* cyst removals show a trend of increasing log removal with increasing spike dose. The regression analysis produced a slope estimate of 0.14 with a standard error of 0.03 and a *p*-value less than 0.01. This *p*-value suggests that the trend of increasing log removal with increasing spike dose was statistically significant at a 95 percent level of confidence. As with the oocyst data, multiple regression was performed to determine if other variables had a significant impact on cyst removal. In this case, log removal of cysts was positively correlated with initial cyst concentration, but negatively correlated with source water NOM concentration as measured by UV absorbance. Comparisons with previously reported data are consistent with the oocyst comparisons described in the previous paragraphs.

Although the data are consistent with theoretical expectation for flocculation and sedimentation processes, it is important to consider other possible explanations for the observed results. For example, Emelko (2001) has noted that log removal estimates are limited by influent concentration when effluent samples are below detection limit and when the detection limit is then used to calculate the log removal. Although this technique may have been used in other studies, Figures 3.5 and 3.6 do not include log removals calculated from detection limits. Therefore, the errors introduced by such calculations are not included in these figures.

Control experiments were not conducted as part of the present study to assess losses to experimental equipment. Previous bench-scale studies at the University of Wisconsin have shown that losses are less than 0.3-log for sedimentation processes (Harrington *et al.* 2001). Additional studies conducted with bench-scale and pilot-scale equipment have shown losses less than 0.25-log for filtration experiments conducted in Waterloo and Ottawa (Emelko 2001). In most cases, the observed differences between high concentration and low concentration treatment trains are larger than these observed losses.

As noted in Chapter 2, analytical recoveries were not used in the calculation of log removal and this could also influence the interpretation of results. However, sensitivity analyses have shown that the inclusion of recovery factors could change the log removal estimates by somewhere between 0.0 and 0.4 logs. For example, in the January 2005 experiment, *Cryptosporidium* oocyst concentrations in the high concentration train were 7.0 x 10^4 oocysts/L and 5.7 x 10^3 oocysts/L for the spiked raw water and the sedimentation effluent, respectively. Using these numbers, the estimated log removal of oocysts in this treatment train was 1.1. With the same approach, the estimated log removal in the low concentration treatment train was 0.1, which is 1.0 log smaller than the estimated log removal in the high concentration train. Recovery experiments conducted with 5 x 10^1 oocysts/L showed median recoveries of 69 percent from raw water and 68 percent from sedimentation effluent. Using these recoveries, adjusted *Cryptosporidium* oocyst concentrations in the high concentration train were 1.0 x 10^5 oocysts/L and 8.4 x 10^3 oocysts/L for the spiked raw water and the sedimentation effluent, respectively. These numbers also produce a log removal estimate of 1.1 and a similar approach produces a log removal estimate of 0.1 log in the low concentration train. Thus, use of these recovery factors had no change on log removal estimates or on the difference between the two treatment trains.

However, as noted in Chapter 2, it may be inappropriate to apply the recovery data to observed data from the high concentration train because the observed oocyst concentration was significantly different from the oocyst concentration used in the recovery experiment. Data from Borchardt and Spencer (2002) suggest that recoveries of 92 percent or more can be obtained at the concentrations observed in the high concentration treatment train. Using a 95 percent value as part of a hypothetical sensitivity analysis, adjusted *Cryptosporidium* oocyst concentrations in the high concentration train would be 7.4 x 10^4 oocysts/L and 6.0 x 10^3 oocysts/L for the spiked raw water and the sedimentation effluent, respectively. These numbers yield a log removal estimate of 1.1, which is identical to the log removal estimates calculated earlier. For the November 2004 and January 2005 experiments, the difference in log removal between the high concentration train and the low concentration train was approximately 1 log, which is likely to be greater than the error introduced by the lack of including recovery factors in the log removal estimate. This is in agreement with the conclusions reached by Dugan *et al.* (2001).

Emelko (2001) has reported that "counting 10 or more oocysts in filter effluent samples will substantially decrease the uncertainty associated with calculations of log removals." In the present study, *Giardia* cysts and *Cryptosporidium* oocysts were detected in 24 of the 36 samples collected from the effluent of a sedimentation basin. With the microscopy method, 10 or more *Giardia* cysts and 10 or more *Cryptosporidium* oocysts were counted in all 24 of these samples. These 24 samples were used to generate the 24 data points shown on Figures 3.5 and 3.6. As noted earlier, results from the FC sorting-microscopy method were not used to estimate concentrations in sedimentation effluent samples. The collection of 10 L grab samples in the present study facilitated the ability to achieve counts of 10 or more organisms from each sample.

Based on this qualitative analysis of uncertainties in the data, it is apparent that the differences in log removal between high and low concentration trains can be attributed to the spike dose used in the treatment train.

PATHOGEN REMOVAL FROM ALUM ADDITION THROUGH GRANULAR MEDIA FILTRATION

In all six experiments, log removals on the high concentration train were higher than log removals on the low concentration train (see Appendices A through F). To determine if general trends could be observed by combining results from all six experiments, the grab and continuous sample results from all six experiments are shown in Figures 3.7 and 3.8 for oocysts and cysts, respectively. These two figures do not show log removals for trains that achieved pathogen concentrations that were below the detection limit. For *Cryptosporidium* oocyst removal, the slope of 0.21 in Figure 3.7 had a standard error of 0.04 and a p-value less than 0.01. A slope of 0.23 was observed for the relationship between *Giardia* cyst removal and the log of the initial spike concentration. This slope had a standard error of 0.04 and a p-value less than 0.01 (Figure 3.8). Because the *Cryptosporidium* and *Giardia* slopes had p-values less than 0.05, the slopes are considered to be significantly different from zero at a confidence level greater than 95 percent.

The side-by-side experimental design of this study clearly showed initial concentration to be an important variable in all six experiments (see Appendices A through F). However, Figures 3.7 and 3.8 illustrate that initial concentration cannot explain all of the variation in log removal. This is to be expected due to changes in source water quality and treatment operations from one experiment to another. As with the sedimentation results, multiple regression was performed on these results to assess whether any of the other measured variables were significant. For *Cryptosporidium* oocysts, log removal was found to correlate with initial oocyst concentration and raw water turbidity. The result suggested that higher log removal was observed at higher raw water turbidity. The same analysis was done for *Giardia* cysts. This result demonstrated that log cyst removal was a function of initial cyst concentration, raw water turbidity and sampling method (either grab or continuous sampling). In this case, continuous sampling produced a higher log cyst removal than grab sampling.

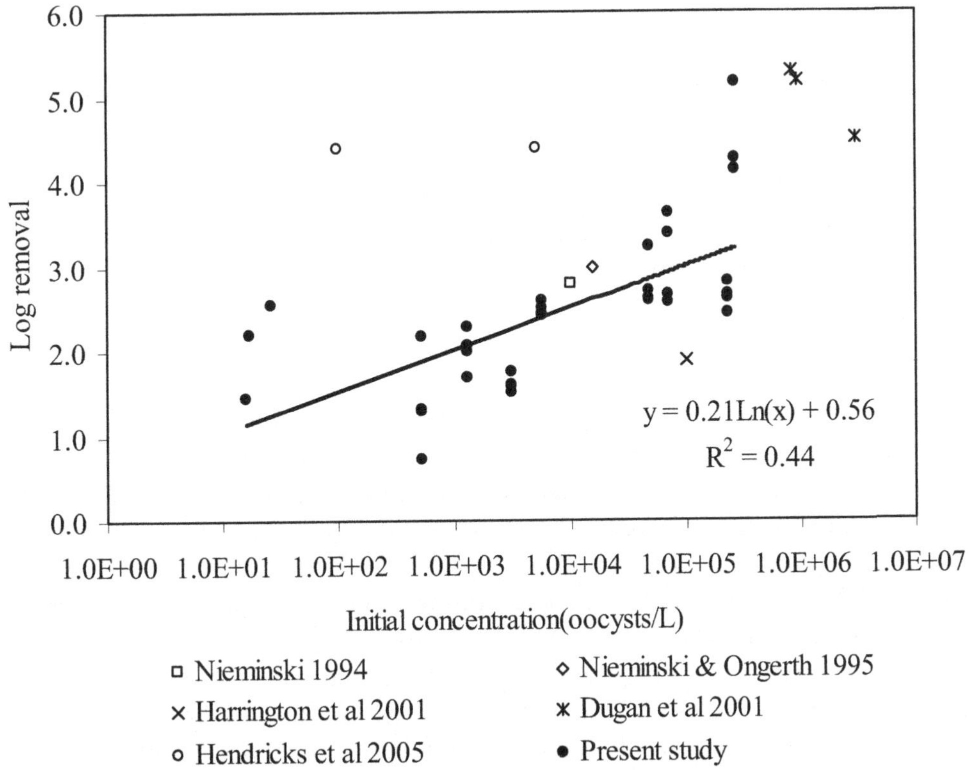

Figure 3.7 ***Cryptosporidium*** **oocyst removal through conventional treatment.**

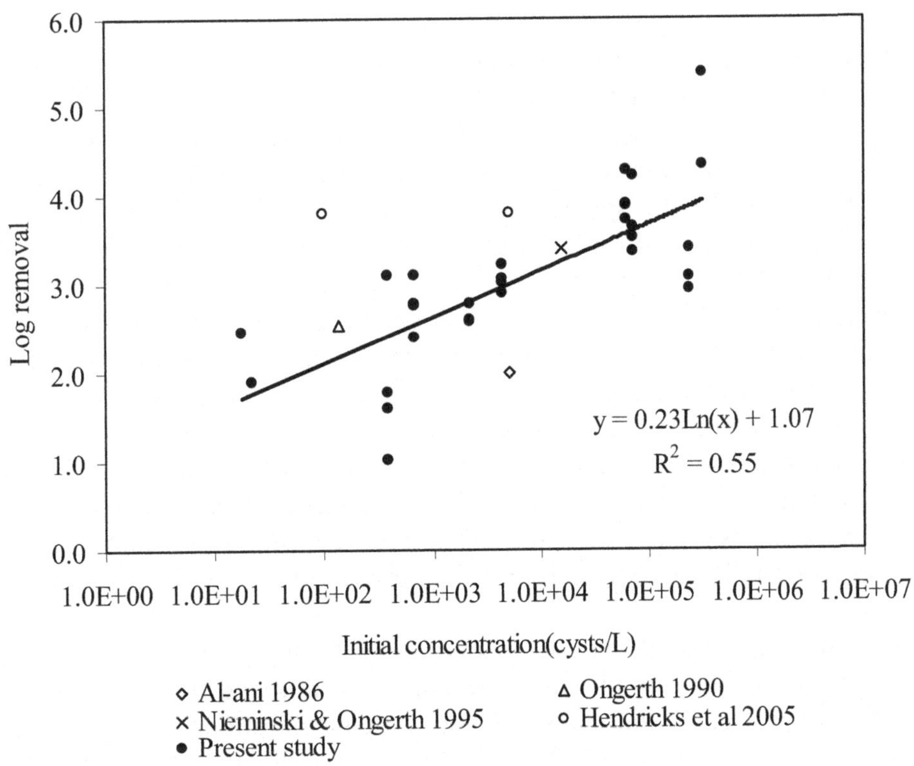

Figure 3.8 ***Giardia*** **cyst removal through conventional treatment.**

In addition, results were also compared with results obtained by other investigators. For *Cryptosporidium* oocysts, pilot scale data have suggested that conventional treatment can achieve oocyst removals of 1 to 5 log (see Figure 3.7). As mentioned earlier, other factors can contribute to the reported differences in *Cryptosporidium* removal capabilities of conventional treatment (e.g., differences in raw water quality, treatment operation, and influent microorganism concentrations). Differences between the present study and those of Dugan *et al.* (2001) were described in the previous section. The study by Hendricks *et al.* (2005) involved the simultaneous spiking of several microorganisms at spike doses of 1 x 10^1 to 5 x 10^3 *Giardia* cysts/L, 1 x 10^1 to 5 x 10^2 *Cryptosporidium* oocysts/L, 5 x 10^4 to 5 x 10^5 algae cells/L, and 1 x 10^7 to 1 x 10^9 bacteria cells/L. Raw water particle counts varied from 1 x 10^6 to 6 x 10^6 particles/L.

Although the data are consistent with theoretical expectation for conventional treatment processes, it is important to consider other possible explanations for the observed results. For example, Emelko (2001) has noted that log removal estimates are limited by influent concentration when effluent samples are below detection limit and when the detection limit is then used to calculate the log removal. Although this technique may have been used in other studies, Figures 3.7 and 3.8 do not include log removals calculated from detection limits. Therefore, the errors introduced by such calculations are not included in these figures.

Control experiments were not conducted as part of the present study to assess losses to experimental equipment. Previous bench-scale studies at the University of Wisconsin have shown that losses are less than 0.3-log for sedimentation processes (Harrington *et al.* 2001). Additional studies conducted with bench-scale and pilot-scale equipment have shown losses less than 0.25-log for filtration experiments conducted in Waterloo and Ottawa (Emelko 2001). In most cases, the observed differences between high concentration and low concentration treatment trains are larger than these observed losses.

As noted in Chapter 2, analytical recoveries were not used in the calculation of log removal and this could also influence the interpretation of results. However, sensitivity analyses have shown that the inclusion of recovery factors could change the log removal estimates by somewhere between 0.0 and 0.4 logs. For example, in the December 2006 experiment, average *Cryptosporidium* oocyst concentrations in the low concentration train were 2.6 x 10^1 oocysts/L and 6.5 x 10^{-2} oocysts/L for the spiked raw water and the filter effluent, respectively. Using these numbers, the estimated log removal of oocysts in this treatment train was 2.6. With the same approach, the estimated log removal in the high concentration treatment train was 5.2, which is twice as high as the estimated log removal in the low concentration train. Recovery experiments conducted with 5 x 10^1 oocysts/L showed median recoveries of 69 percent from raw water and 63 percent from filter effluent. Using these recoveries, adjusted *Cryptosporidium* oocyst concentrations in the low concentration train were 3.8 x 10^1 oocysts/L and 1.0 x 10^{-1} oocysts/L for the spiked raw water and the sedimentation effluent, respectively. These numbers also produce a log removal estimate of 2.6 for the low concentration train and a similar approach produces a log removal estimate of 5.2 log in the high concentration train. Thus, use of these recovery factors had no change on log removal estimates or on the difference between the two treatment trains.

As noted in Chapter 2, it may be inappropriate to apply the recovery data to observed data from the spiked raw water in the high concentration train and from the filter effluent in the low concentration train because these observed oocyst concentrations were significantly different from the oocyst concentration used in the recovery experiment. Data from Borchardt and

Spencer (2002) suggest that recoveries of 92 percent or more can be obtained at the concentrations observed in the spiked raw water on the high concentration treatment train. No data have been published to illustrate recoveries obtained at the concentrations observed in the filter effluent on the low concentration train. To assess the potential magnitude of recovery factor on log removal estimates, a hypothetical recovery of 95 percent was used for the spiked raw water on the high concentration train and a hypothetical recovery of 25 percent was used for the filter effluent on the low concentration train. In this sensitivity analysis, a recovery factor of 69 percent was used for the spiked raw water on the low concentration train and a recovery factor of 63 percent was used for the filtered water on the high concentration train. With these assumed recovery factors, log removal estimates were 2.2 on the low concentration train and 5.0 on the high concentration train. Thus, the log removal estimate was decreased by 0.4 for the low concentration train and by 0.2 for the high concentration train. The difference of 2.8 log between the two trains was 0.2 log higher than the difference of 2.6 that was estimated without recovery factors. Similar sensitivity analyses with other experiments suggest that the difference in log removal between the two treatment trains is likely to be greater than the error introduced by the lack of including recovery factors in the log removal estimate. This is in agreement with the conclusions reached by Dugan *et al.* (2001).

Emelko (2001) has reported that "counting 10 or more oocysts in filter effluent samples will substantially decrease the uncertainty associated with calculations of log removals." In the present study, *Cryptosporidium* oocysts were detected in 11 of 12 continuous samples and in 23 of 36 grab samples collected from the effluent of a filter. *Giardia* cysts were detected in 10 of 12 continuous samples and in 22 of 36 grab samples. These samples with detectable concentrations were used to generate the 34 data points shown on Figure 3.7 and the 32 data points shown on Figure 3.8. With the microscopy method, 10 or more *Giardia* cysts were counted in 25 of 32 samples with detectable concentrations and 10 or more *Cryptosporidium* oocysts were counted in 30 of 34 samples with detectable concentrations. With the FC sorting-microscopy method, 10 or more *Giardia* cysts were counted in 19 of 32 samples with detectable concentrations and 10 or more *Cryptosporidium* oocysts were counted in 24 of 34 samples with detectable concentrations. The collection of 10 L grab samples and 288 L continuous samples in the present study facilitated the ability to achieve counts of 10 or more organisms from most samples.

PATHOGEN REMOVAL FROM SEDIMENTATION THROUGH GRANULAR MEDIA FILTRATION

As shown in Appendices A and B, the November 2004 and January 2005 experiments did not demonstrate an effect of filter influent pathogen concentration on log removal through the unit process of granular media filtration. For the four remaining experiments, grab samples collected from the sedimentation basin effluent were below the detection limit in the treatment train with the lower initial pathogen concentration (see Appendices C through F). For these same experiments, log removals ranged from 0.3 to 4.0 in the treatment train with the higher initial pathogen concentration. However, because log removals could not be calculated for the low concentration trains, a direct comparison of performance between the two trains was not feasible in the latter four experiments.

For all six experiments, results from sedimentation effluent to filtration effluent were used to generate Figures 3.9 and 3.10, which show a summary of *Cryptosporidium* and *Giardia* results, respectively. As with the previous sections, these two graphs do not show data from

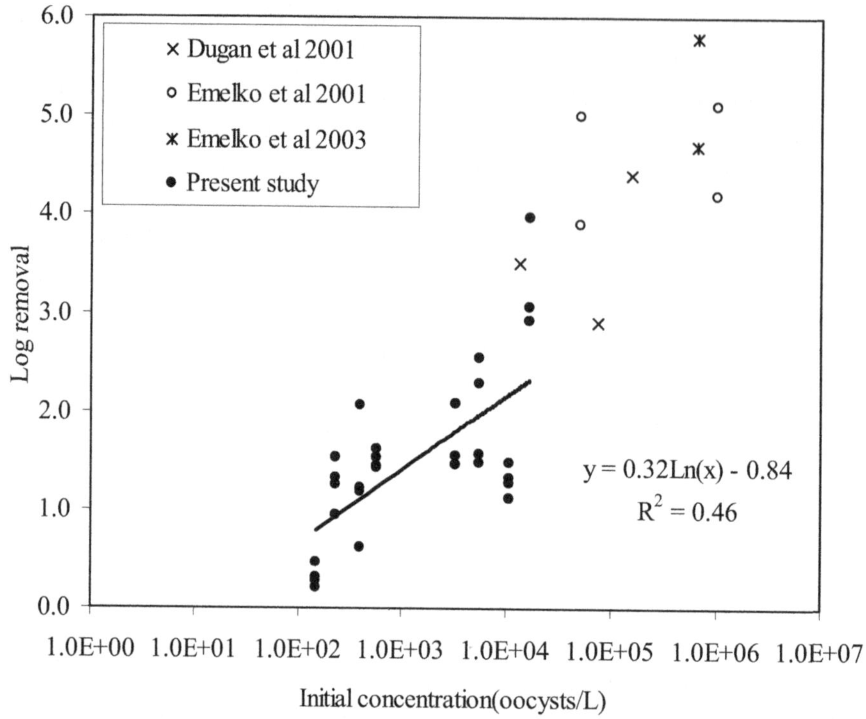

Figure 3.9 ***Cryptosporidium*** **oocyst removal through filtration.**

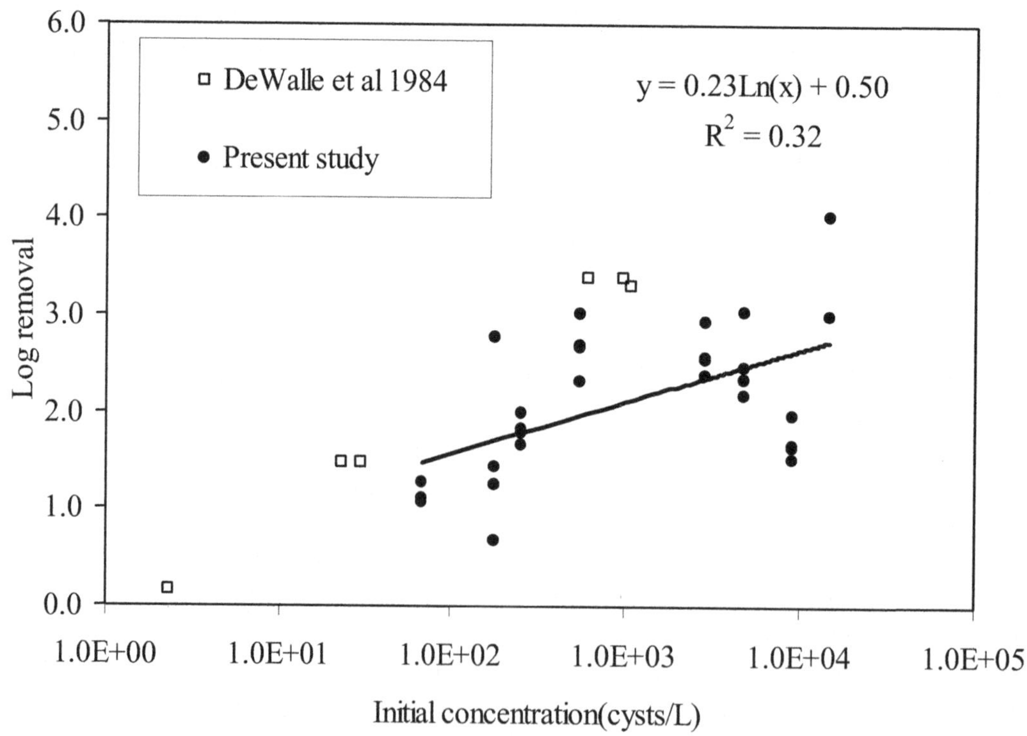

Figure 3.10 ***Giardia*** **cyst removal through filtration.**

treatment trains with concentrations below the detection limit. Results obtained by other investigators are also shown. For oocyst removal, regression analysis estimated a slope of 0.32 with a standard error of 0.07 and a p-value less than 0.01. The same analysis was performed for the *Giardia* cyst results (Figure 3.9). *Giardia* cyst removal was correlated with filter influent concentration with a slope of 0.23. This slope had a standard error of 0.06 and a p-value less than 0.01. Because the *Cryptosporidium* and *Giardia* slopes had a p-value less than 0.05, the observed slope can be considered to be significantly different from zero at a confidence level greater than 95 percent. Results from other studies appear to be qualitatively consistent with the trend of increasing log removal with increasing filter influent concentration of microorganisms.

Although the data are consistent with previous filtration studies, it is important to consider other possible explanations for the observed results. The previous section described the possible impacts of system losses, detection limits, analytical recoveries, and pathogen counts on the interpretation of results through conventional treatment. That discussion also applies to the interpretation of results through the filtration unit process.

COMPARISON OF GRAB SAMPLING WITH CONTINUOUS SAMPLING

Figures 3.11 and 3.12 show a comparison of mean log removals based on grab sampling with those based on continuous sampling. Error bars in these figures show the standard

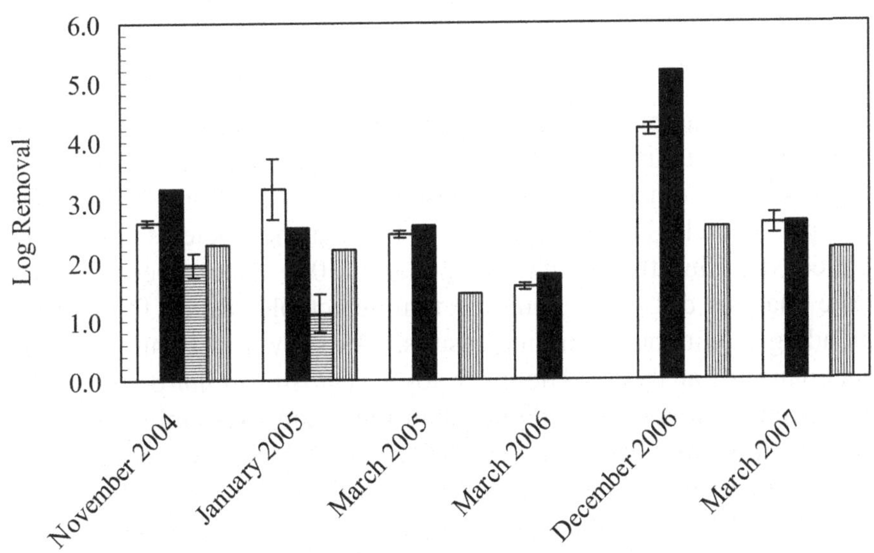

Figure 3.11 ***Cryptosporidium*** **oocyst removal through conventional treatment, analyzed by microscopy (some filter effluent grab samples collected during the March 2005, March 2006, December 2006, and March 2007 experiments had no detectable pathogens in them, and average log removals could not be calculated for these samples). Symbols: □ Grab Sampling with High Concentration, ■ Continuous sampling with High Concentration, ⊟ Grab sampling with Low Concentration, ▥ Continuous Sampling with Low Concentration**

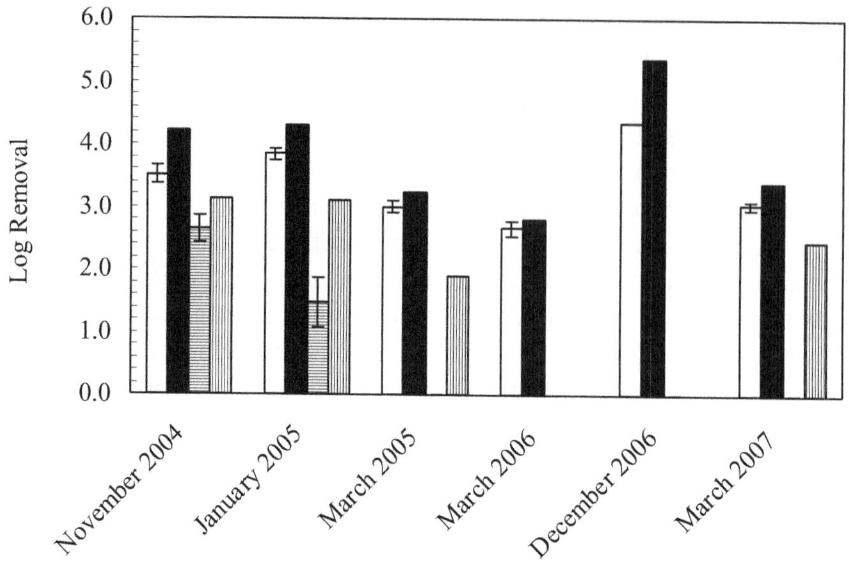

Figure 3.12 ***Giardia*** **cyst removal through conventional treatment analyzed by microscopy (some filter effluent grab samples collected during the March 2005, March 2006, December 2006 and March 2007 experiments had no detectable pathogens in them, and average log removals could not be calculated for these samples). Symbols: □ Grab Sampling with High Concentration, ■ Continuous sampling with High Concentration, ⊟ Grab sampling with Low Concentration, ◫ Continuous Sampling with Low Concentration.**

deviation observed among sampling times for grab samples. Paired *t*-tests were run to determine if there was a statistically significant difference between the log removals obtained from grab samples and log removals obtained from continuous samples for *Cryptosporidium* oocysts. Spike doses of approximately 10^1 pathogens/L resulted in non-detects for many filter effluent grab samples collected during the March 2005, March 2006, December 2006 and March 2007 experiments. The analysis did not include the non-detectable results from these grab samples and the corresponding continuous sampling results. As shown in Figure 3.11, this produced a total of eight data pairs for analysis. The mean log *Cryptosporidium* oocyst removal calculated from continuous sampling was 0.3-log higher than the mean log removal calculated from grab sampling. However, the *t*-test demonstrated that the difference between grab and continuous sampling was not statistically significant with a *p*-value of 0.12. The same analysis was performed on the eight data pairs obtained for *Giardia* cysts (see Figure 3.12). The mean log *Giardia* cyst removal calculated from continuous sampling was 0.6-log higher than the mean log pathogen removal calculated from grab sampling and this difference was significant at a 95% confidence level (p-value < 0.01).

CRYPTOSPORIDIUM OOCYST REMOVAL VERSUS *GIARDIA* CYST REMOVAL

The results obtained from all six experiments were used to generate Figures 3.13 to 3.17. As in previous sections, error bars show the standard deviation observed among sampling times. Figure 3.13 compares log removal of *Cryptosporidium* oocysts with log removal of *Giardia* cysts through sedimentation. A paired *t*-test demonstrated that the difference in log removal was

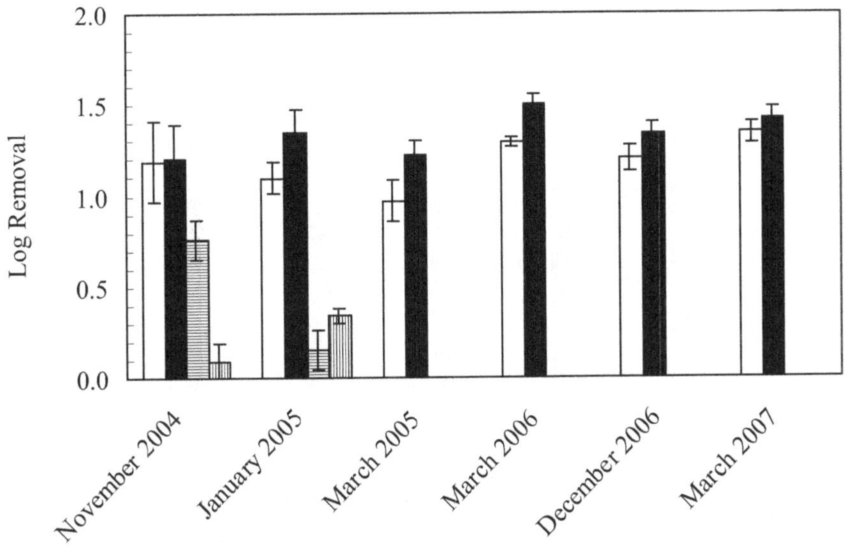

Figure 3.13 Pathogen removals through sedimentation (some sedimentation effluent samples collected during the March 2005, March 2006, December 2006 and March 2007 experiments had no detectable pathogens in them, and average log removals could not be calculated for these samples). Symbols: □ *Cryptosporidium* with High Concentration, ■ *Giardia* with High Concentration, ⊟ *Cryptosporidium* with Low Concentration, ◫ *Giardia* with Low Concentration.

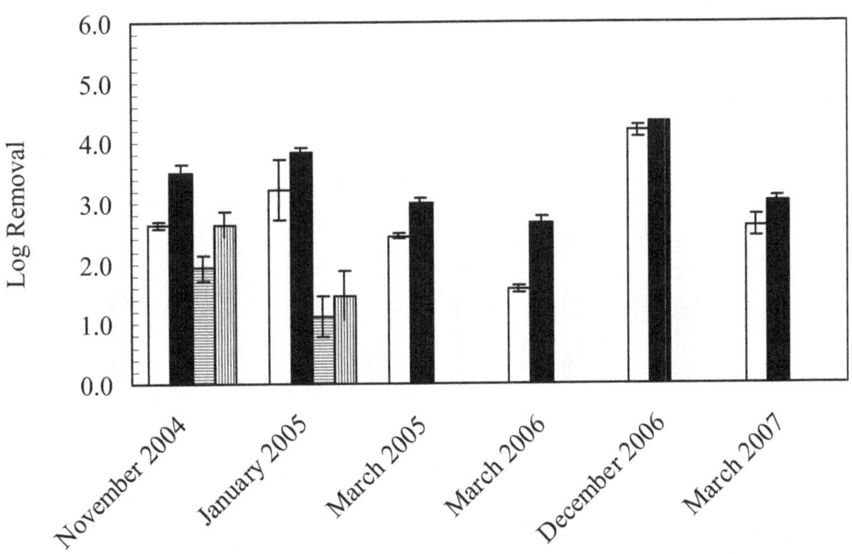

Figure 3.14 Pathogen removals through conventional treatment, based on grab samples (some filter effluent samples collected during the March 2005, March 2006, December 2006 and March 2007 experiments had no detectable pathogens in them, and average log removals could not be calculated for these samples). Symbols: □ *Cryptosporidium* with High Concentration, ■ *Giardia* with High Concentration, ⊟ *Cryptosporidium* with Low Concentration, ◫ *Giardia* with Low Concentration.

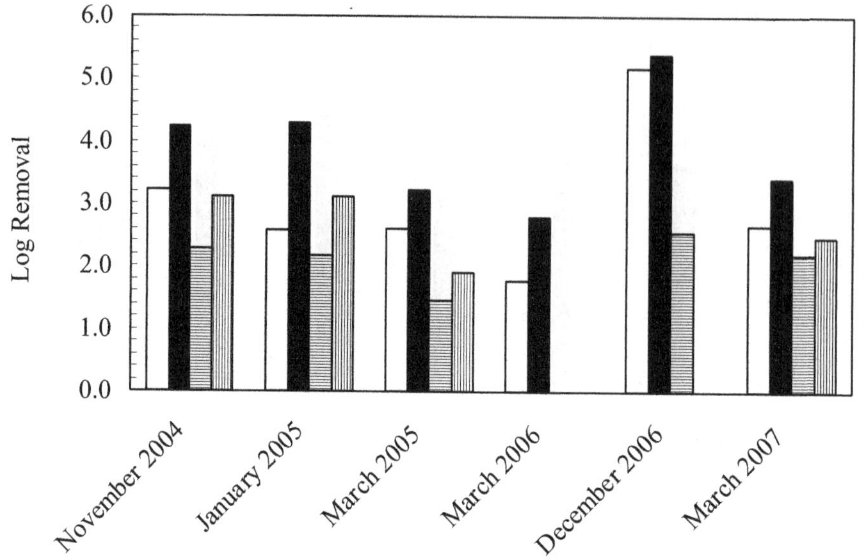

Figure 3.15 **Pathogen removals through conventional treatment, based on continuous samples (filter effluent samples at low concentration collected during the March 2006 experiment had no detectable pathogens in them, and log removals could not be calculated for these samples). Symbols: □ *Cryptosporidium* with High Concentration, ■ *Giardia* with High Concentration, ⊟ *Cryptosporidium* with Low Concentration, ▥ *Giardia* with Low Concentration.**

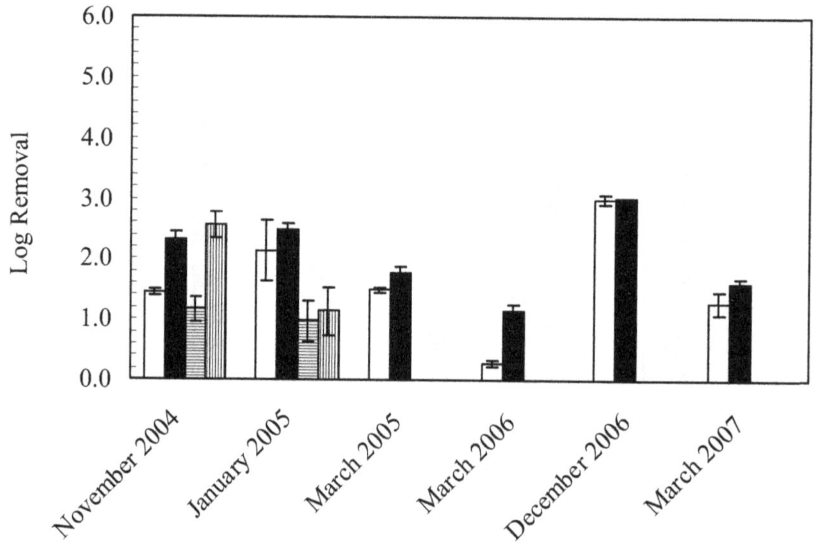

Figure 3.16 **Pathogen removals through filtration, based on grab samples (some samples collected during the March 2005, March 2006, December 2006 and March 2007 experiments had no detectable pathogens in them, and average log removals could not be calculated for these samples). Symbols: □ *Cryptosporidium* with High Concentration, ■ *Giardia* with High Concentration, ⊟ *Cryptosporidium* with Low Concentration, ▥ *Giardia* with Low Concentration.**

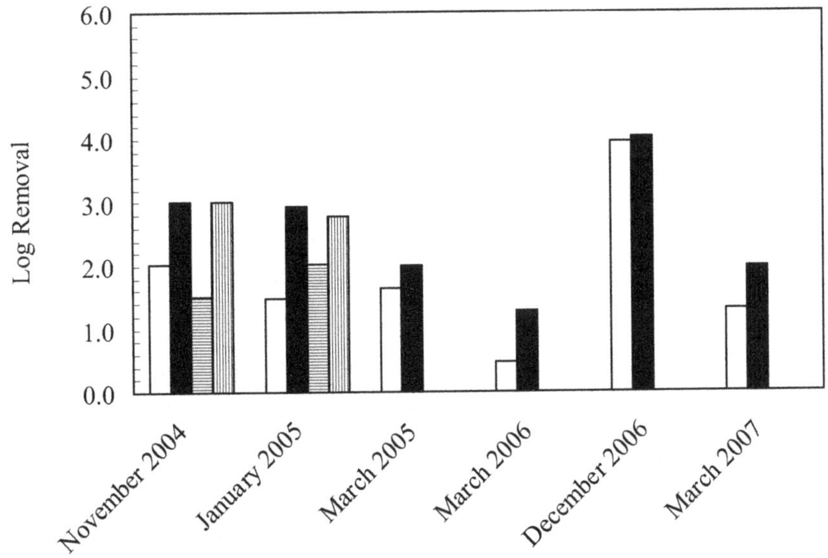

Figure 3.17 Pathogen removals through filtration, based on continuous samples (some samples collected during the March 2005, March 2006 and December 2006 experiments had no detectable pathogens in them, and log removals could not be calculated for these samples). Symbols: □ *Cryptosporidium* with High Concentration, ■ *Giardia* with High Concentration, ⊟ *Cryptosporidium* with Low Concentration, ◫ *Giardia* with Low Concentration.

significant at a 95% confidence level (p-value < 0.01). However, the mean log *Giardia* cyst removal was only 0.2 log higher than the mean log *Cryptosporidium* oocyst removal. This difference is smaller than the experimental error of the pilot plant. This agreed with a previous study, where slightly better removals of *Giardia* were observed when compared with *Cryptosporidium* but the differences were within the reported variation (Edzwald *et al.* 2000). Figures 3.14 and 3.15 compare log removal of *Cryptosporidium* oocysts with log removal of *Giardia* cysts through conventional treatment. In this case, a paired t-test showed that *Giardia* cysts were more readily removed than *Cryptosporidium* oocysts (p-value < 0.01). The mean log *Giardia* cyst removal was 0.7 log higher than the mean log *Cryptosporidium* oocyst removal. The higher level of *Giardia* cyst removal may be related to its relatively larger particle size. This conclusion agrees with findings obtained by previous studies that have generally found *Giardia* cyst removals on the order of 0.3 to 1 log higher than removals of *Cryptosporidium* oocysts (Nieminski 1994, Nieminski and Ongerth 1995, States *et al.* 1997, and Edzwald *et al.* 2000). Figures 3.16 and 3.17 compare log removal of *Cryptosporidium* oocysts with log removal of *Giardia* cysts through filtration. A paired t-test showed that *Giardia* cysts were more readily removed than *Cryptosporidium* oocysts within this single unit process (p-value < 0.01). Similar to the conventional treatment train results, the mean log *Giardia* cyst removal was 0.6 log higher than the mean log *Cryptosporidium* oocyst removal. This is consistent with the expectation that cyst-sized particles will be removed more effectively than oocyst-sized particles according to filtration theory as described by Amirtharajah (1988) and by Tobiason and O'Melia (1988).

CHAPTER 4
SUMMARY AND CONCLUSIONS

Side-by-side treatment trains were used to test the significance of pathogen spike dose on pathogen removal in pilot-scale conventional treatment. The following items summarize the conclusions obtained from this study.

1. The removal of *Cryptosporidium* oocysts and *Giardia* cysts through conventional treatment was observed to depend on initial pathogen spike dose. Removal increased with increasing spike dose and statistical analyses suggested that log removal differences between low and high concentration trains were statistically significant in most cases. As described in Chapter 1, this is consistent with theoretical expectation.
2. As with conventional treatment, removal of *Cryptosporidium* oocysts and *Giardia* cysts by the unit process of granular media filtration appeared to correlate with pathogen concentration at the filter influent. Lower pathogen removals were observed at lower pathogen concentrations.
3. When comparing previously published data with the results of this study, it is apparent that conventional treatment can achieve removals of 1 to 5 log for either *Cryptosporidium* oocysts or *Giardia* cysts. Differences in raw water quality, treatment operation, and influent microorganism concentrations can all contribute to the reported differences in the pathogen removal capabilities of conventional treatment.
4. Continuous sampling allowed for the calculation of log removals in most cases when grab sampling produced non-detectable concentrations of *Cryptosporidium* oocysts or *Giardia* cysts. This approach provided an opportunity to evaluate pathogen spike doses of approximately 10^1 pathogens/L. For the 8 cases in which both continuous sampling and grab sampling produced detectable concentrations, calculations from continuous samples produced a higher mean log removal of *Giardia* cysts than did calculations from grab samples for *Giardia* cysts. However, this was not observed for *Cryptosporidium* oocysts.
5. For raw water and filtered water samples, the results obtained by microscopy were not significantly different from those obtained by FC sorting-microscopy. However, settled water samples could not be sorted by the flow cytometer due to clogging of the flow cell. Therefore, FC sorting-microscopy appears to be a useful method for samples with relatively low particle content.
6. Mean log removal of *Giardia* cysts was significantly greater than mean log removal of *Cryptosporidium* oocysts through conventional treatment and through the filtration step. However, there was little difference between the two pathogens with respect to log removal through sedimentation.

CHAPTER 5
RECOMMENDATIONS TO UTILITIES

The first of the conclusions described in the previous chapter has important implications for the drinking water community. The USEPA currently allows a 3-log credit for *Cryptosporidium* oocyst removal in conventional surface water treatment plants. For the water supply evaluated in this study, this level of removal was only observed at initial oocyst concentrations that were many orders of magnitude higher than typically observed in a drinking water supply. Thus, it is plausible that some surface water systems may not achieve the level of *Cryptosporidium* oocyst removal they are given credit for.

These conclusions were based on experiments conducted at relatively low temperatures and with a water supply that has a relatively low particle content and a moderate dissolved organic carbon content. Similar experiments conducted on other water supplies may be expected to produce different conclusions. For example, sedimentation and filtration performance can both be expected to depend on total particle content and its influence on aggregation kinetics. Total particle content depends on raw water particle content, pathogen spike dose, and the amount of aluminum hydroxide particles created by alum addition. In this study, variations in pathogen spike dose produced significant variations in total particle content. For example, a pathogen spike of 10^6 pathogens/L changed the influent total particle content from approximately 1×10^6 particles/L to 2×10^6 particles/L. Therefore, for the water evaluated in this study, the high pathogen spike of 10^6 pathogens/L increased the influent particle concentration by a factor of 2. However, national average source water particle concentrations of 22,000/mL (2.2×10^7 particles/L) have been reported in previous AwwaRF-funded research (McTigue *et al.* 1998). In pilot-scale investigations of waters like these, a pathogen spike of 10^6 pathogens/L will only change the influent total particle concentration from 2.2×10^7 to 2.3×10^7 particles/L. Thus, the spiked water would have a total particle concentration that is only 1.05 times higher than the typically observed particle concentration. Because of the site-specific nature of particle content, water utilities are encouraged to participate in research studies that elucidate the impacts of overall particle content and particle characteristics on pathogen removal in conventional treatment processes.

Temperature is also known to influence the rate of particle aggregation in flocculation and sedimentation processes. The results obtained in this study were derived from relatively cold water conditions of 7 to 15°C, which are lower temperatures than have been used in previous studies. The drinking water industry is encouraged to participate in more extensive studies designed to evaluate the impact of temperature on pathogen removal in conventional drinking water treatment.

The results were also derived from experiments that used alum as the sole coagulant. No attempt was made to enhance particle separation through the use of coagulation aids or filtration aids. In addition, the impacts of preoxidation on treatment process performance were not evaluated. The impact of these operational conditions on the results obtained from this study need to be evaluated.

This research also has some implications for water utilities that are served by sources with significant fluctuations in influent particle load. These fluctuations may be due to seasonal events such as snow melt, periodic storm events, or to hydrologic conditions (e.g., river sources versus lake sources). The results suggest that pathogen removals may actually improve during periods of elevated turbidity and/or particle load, provided that optimal coagulation conditions are maintained at all times. However, utilities are cautioned to recognize that this will not guarantee a lower filter effluent concentration (see calculations in the next paragraph). As always, utilities are strongly encouraged to maintain optimal operational conditions and to be vigilant in monitoring of filter effluent water quality.

The water industry is also encouraged to participate in risk assessments of the conclusions presented in this report. For example, studies with our pilot plant show an estimated 3-log removal of *Cryptosporidium* oocysts with a spike dose of $10^{5.5}$ oocysts/L, suggesting an average filter effluent concentration of $10^{2.5}$ oocysts/L. Meanwhile, an estimated 1-log removal is expected with a spike dose of 10^{1} oocysts/L, suggesting an average filter effluent concentration of 10^{0} oocysts/L. This latter filter effluent concentration is not as low as the 10^{-2} oocysts/L predicted by a 3-log removal credit. Therefore, a reassessment of public health risks is warranted, both now and after completion of studies recommended in the above paragraph.

Continuous separation channel centrifugation was shown to facilitate the understanding of pathogen removal at concentrations closer to the realistic levels that occur in natural water. Because the instrument allows for processing large volumes of filtered water, it can substantially reduce the uncertainty of calculating log removal due to low pathogen concentrations. Utilities are encouraged to work with manufacturers of these devices to assess the possibility of using them for monitoring of conventional treatment performance.

APPENDIX A
EXPERIMENTAL CONDITIONS AND RESULTS FOR THE NOVEMBER 2004 EXPERIMENT

An alum dose of 80 mg/L was set for the treatment train (Train 1) with initial concentrations of 4.6 x 10^4 ± 1.4 x 10^4 *Cryptosporidium* oocysts/L and 7.0 x 10^4 ± 1.5 x 10^4 *Giardia* cysts/L. The other treatment train (Train 2), spiked with 1.3 x 10^3 ± 1.3 x 10^3 *Cryptosporidium* oocysts/L and 6.8 x 10^2 ± 9.7 x 10^1 *Giardia* cysts/L, had an alum dose of 70 mg/L. The alum dose was based on jar tests where the source water was supplemented with test pathogens. This was done to ensure that the selected pilot-scale coagulant dose accounted for the concentration of the pathogen feed. Therefore, both treatment trains were run with optimal coagulation conditions. Raw water temperature during the experiment was 15.2 ± 0.5°C. Spiked raw water turbidity averaged 1.58 ± 0.05 ntu for Train 1 and 1.63 ± 0.10 ntu for Train 2 during the experiment.

Pathogen Removal from Alum Addition through Sedimentation

Turbidity Results

Turbidity results in the sedimentation effluent are shown in Figure A.1. An average of 1.13 ± 0.08 ntu was obtained for Train 1 and 1.02 ± 0.11 ntu was obtained for Train 2. A paired *t*-test showed that the two treatment trains had significantly different turbidity values in the

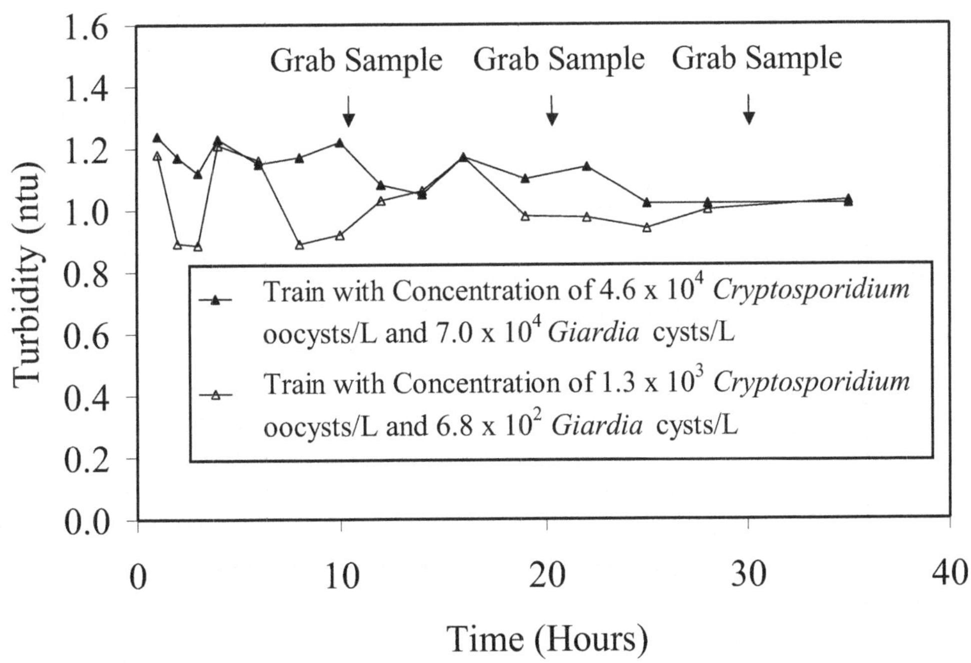

Figure A.1 Turbidity in sedimentation effluent versus run time in the November 2004 experiment.

sedimentation effluent (p-value = 0.004). Train 1 had a significantly higher initial particle concentration than Train 2 because of the higher pathogen dose and because the higher alum dose would have created more aluminum hydroxide particles. This may have led to the higher settled water turbidity in Train 1. For both treatment trains, treatment reduced UV254 by 50 percent from 0.12 cm^{-1} to 0.06 cm^{-1}.

Pathogen Removals through Sedimentation from Grab Samples

Figures A.2 and A.3 illustrate the effect of spike dose on pathogen removal through sedimentation. The figures represent the averaged results obtained from both microscopy and FC sorting-microscopy. With an initial *Cryptosporidium* oocyst concentration of 4.6 x 10^4 oocysts/L, an average of 1.2 ± 0.2 log oocyst removal was achieved by the sequence of coagulation, flocculation, and sedimentation (Figure A.2). The other treatment train, which ran at the lower initial *Cryptosporidium* oocyst concentration of 1.3 x 10^3 oocysts/L, achieved 0.8 ± 0.1 log removal (also see Figure A.2). The same pattern was observed for *Giardia* cyst removal. Treatment trains spiked with *Giardia* cysts at 7.0 x 10^4 cysts/L and 6.8 x 10^2 cysts/L were able to achieve an average log removal of 1.2 ± 0.2 and 0.1 ± 0.1, respectively (Figure A.3). The pathogen removals were about half log to one log higher at higher spike doses than at lower spike doses. A paired t-test showed that the mean log removals in the train with the higher *Giardia* cyst concentration was significantly different at a 95% confidence level from the mean log removals in the train with the lower concentration (p-value < 0.05). The same analysis was

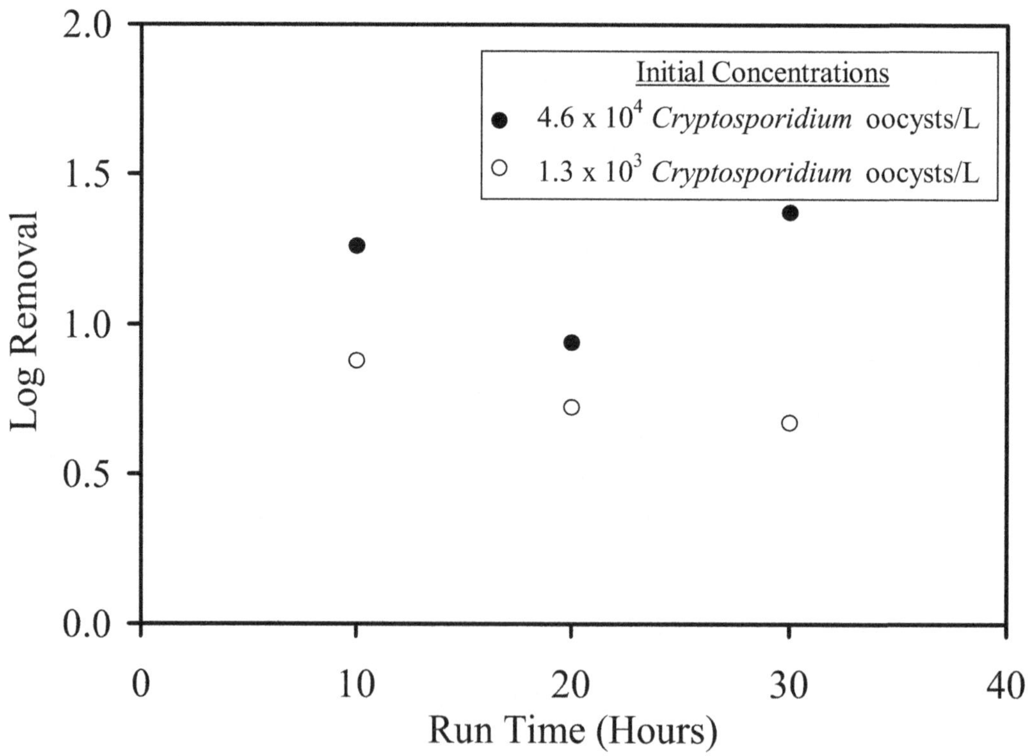

Figure A.2 ***Cryptosporidium*** **oocyst removal through sedimentation based on grab sampling.**

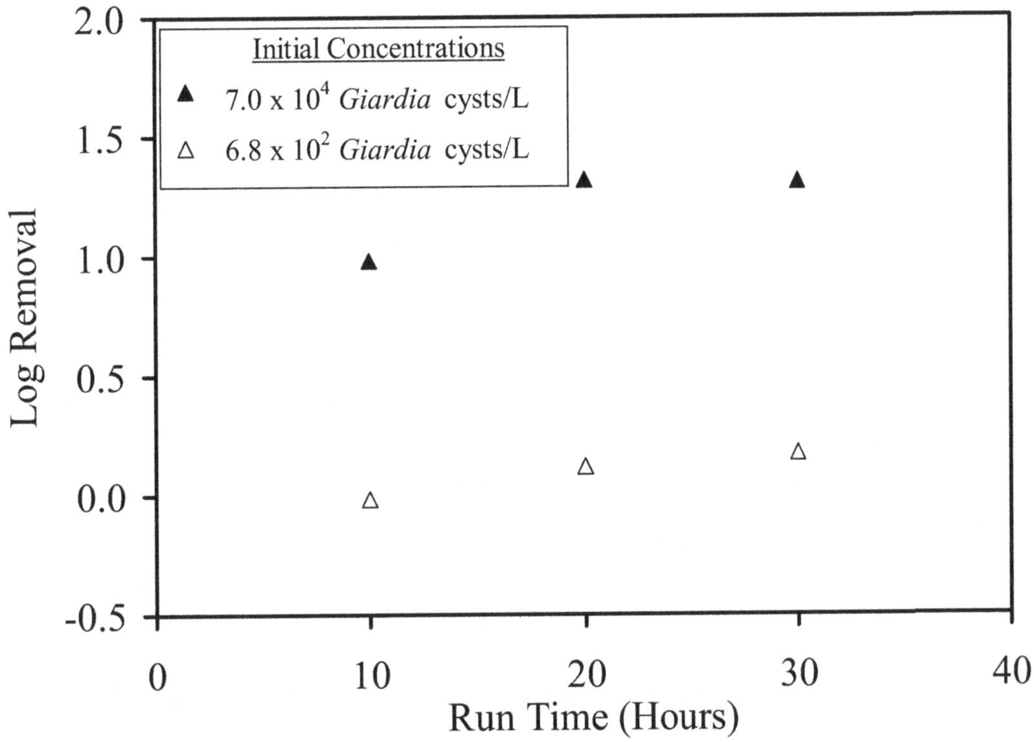

Figure A.3 ***Giardia*** **cyst removal through sedimentation based on grab sampling.**

performed for *Cryptosporidium* oocysts. In this case, the mean log removal in the train with the higher concentration was not significantly different from the mean log removal in the train with the lower concentration (p-value = 0.09).

Pathogen Removal from Alum Addition through Granular Media Filtration

Turbidity Results

During stable operation (as shown in Figure A.4), the average filter effluent turbidity was 0.17 ± 0.03 ntu for the train with the higher pathogen dose and was 0.16 ± 0.02 ntu for the train with the lower pathogen dose. The paired t-test showed that the mean filter effluent turbidity of the train with the higher pathogen dose was not significantly different from the mean filter effluent turbidity of the train with the lower pathogen dose (p-value = 0.09).

Pathogen Removal through Conventional Treatment from Filter Effluent Grab Samples

The effect of initial concentration on the performance of conventional treatment, including coagulation, flocculation, sedimentation and filtration, is shown in Figures A.5 through A.6. The results in these two figures show the averaged log removals calculated from both microscopy and FC sorting-microscopy analyses. When initial concentrations of 4.6 x 10^4 oocysts/L and 7.0 x 10^4 cysts/L were fed into the treatment train, log removals averaged 2.6 ± 0.1 and 3.5 ± 0.1, respectively. As expected, lower log removals were observed with the train

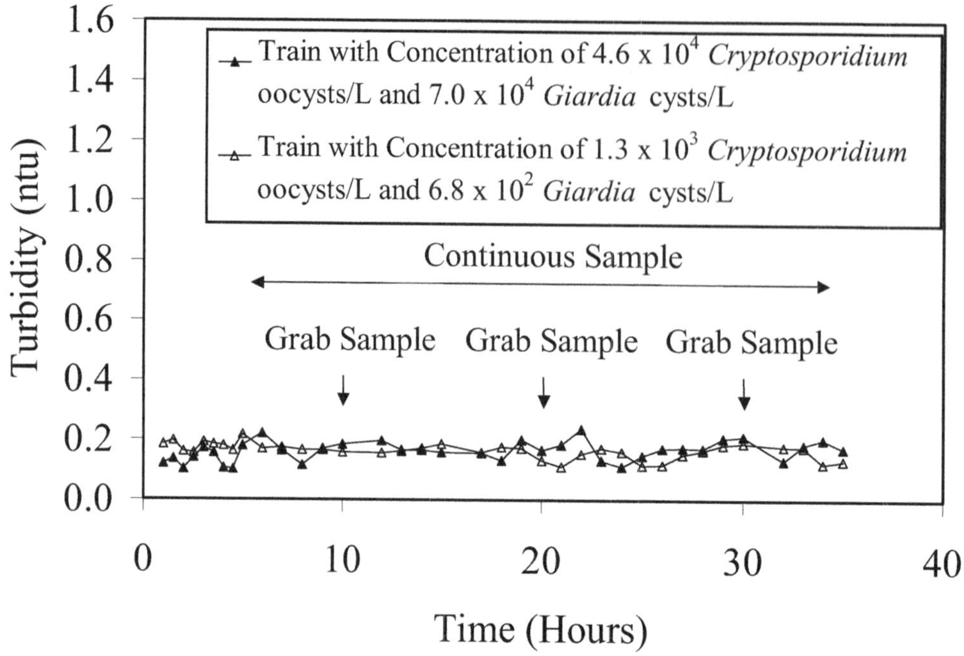

Figure A.4 Turbidity in filtration effluent versus run time in the November 2004 experiment.

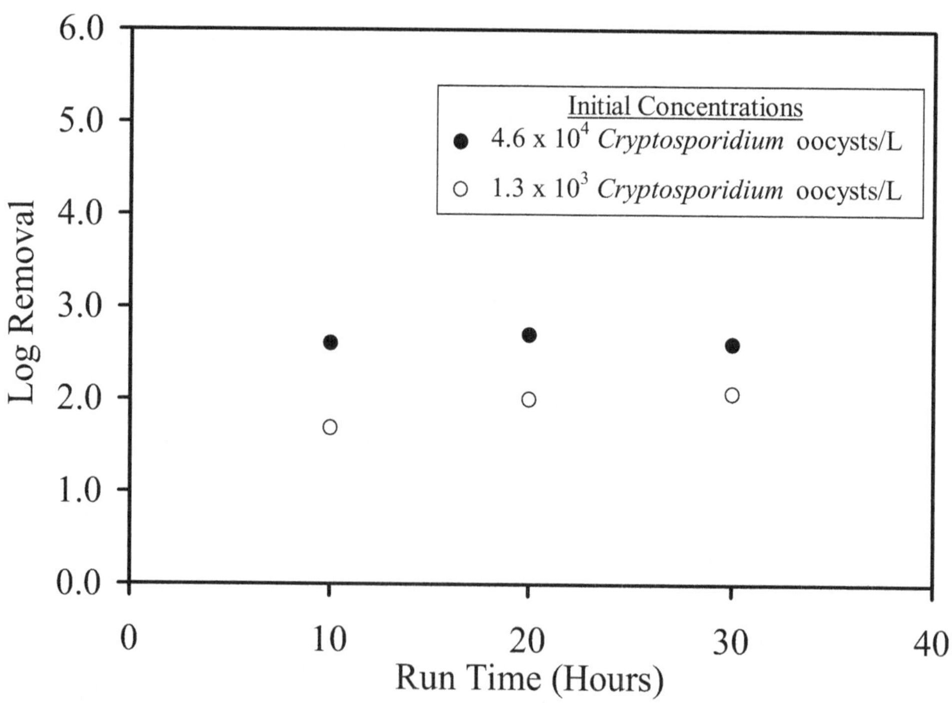

Figure A.5 ***Cryptosporidium*** **oocyst removal through conventional treatment (based on filter effluent grab samples).**

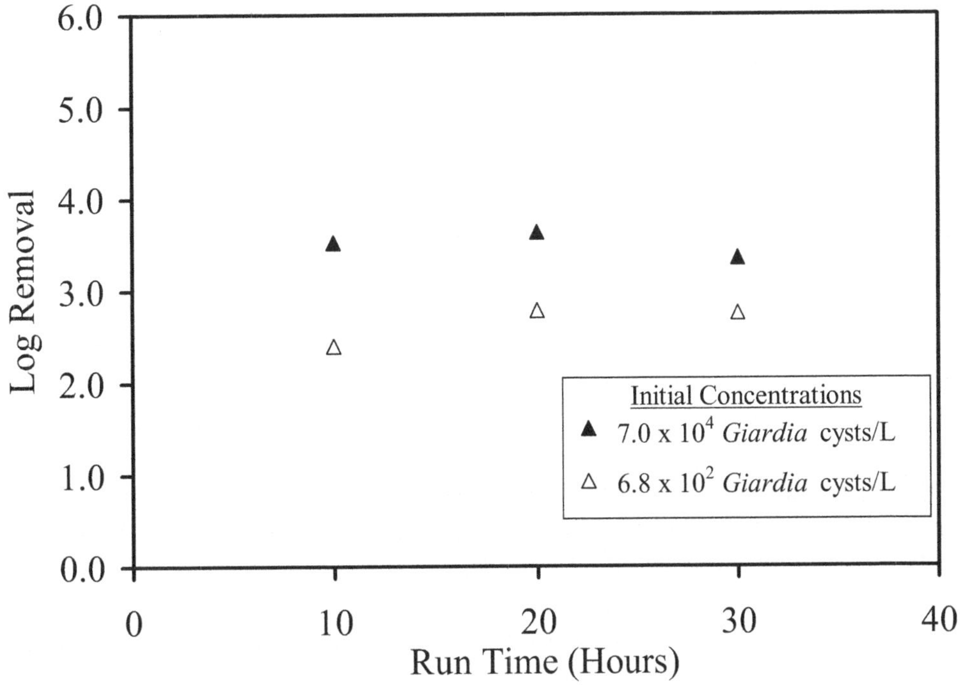

Figure A.6 ***Giardia*** **cyst removal through conventional treatment (based on filter effluent grab samples).**

having the initial concentrations of 1.3 x 10^3 oocysts/L and 6.8 x 10^2 cysts/L. The removals of *Cryptosporidium* oocysts and *Giardia* cysts averaged 1.9 ± 0.2 log and 2.7 ± 0.2 log, respectively. Removal was about one order of magnitude lower at lower spike doses than at higher spike doses for both *Cryptosporidium* oocysts and *Giardia* cysts. A paired *t*-test was run to evaluate the difference between the mean log removals from the two treatment trains. The result suggested the difference between the mean log removals from the two treatment trains was statistically significant for both microbes (p-value = 0.03).

Pathogen Removal through Conventional Treatment for Continuous Samples Collected from Filter Effluent

Log removal results from continuous samples were taken continuously over 30 hours once the ripening phase was complete. The sample times were noted on Figure A.4. For samples with higher spike doses, log removals of 3.2 and 4.2 were observed for *Cryptosporidium* oocysts and *Giardia* cysts, respectively. As expected, removal was lower at lower spike doses (1.3 x 10^3 oocysts/L and 6.8 x 10^2 cysts/L) than at higher spike doses (4.6 x 10^4 oocysts/L and 7.0 x 10^4 cysts/L). Conventional treatment was able to remove 2.3 log and 3.1 log of *Cryptosporidium* oocysts and *Giardia* cysts, respectively, when the lower spike doses were used. The removal was about one log lower at the lower spike doses than at the higher spike doses for both *Cryptosporidium* oocysts and *Giardia* cysts.

Pathogen Removal from Sedimentation through Granular Media Filtration

Pathogen Removal through filtration from Filter Effluent Grab Samples

Log removal results from grab samples are shown in Figures A.7 and A.8 for spiked Lake Mendota water that went from sedimentation to filtration as part of the conventional treatment train. In terms of average *Cryptosporidium* oocyst removal, filtration achieved 1.6 ± 0.3 log and 1.2 ± 0.2 log when the respective pathogen concentrations at the sedimentation basin effluent were 3.3 x 10^3 oocysts/L and 2.3 x 10^2 oocysts/L (see Figure A.7). Average *Giardia* cyst log removals of 2.5 ± 0.4 and 2.6 ± 0.2 were obtained in the treatment trains with settled water *Giardia* cyst concentrations of 4.7 x 10^3 cysts/L and 5.6 x 10^2 cysts/L, respectively (see Figure A.8). A paired *t*-test showed that the log removals in the train with the higher oocyst concentration were not significantly different at a 95% confidence level from the log removals in the train with the lower oocyst concentration (*p*-value = 0.12). Like *Cryptosporidium* oocysts, the same analysis indicated that there was not a significant difference in log cyst removals between the two treatment trains (*p*-value = 0.23).

Pathogen Removal through filtration for Continuous Samples Collected from Filter Effluent

Log removal results from continuous samples were obtained from spiked Lake Mendota water that went from sedimentation to filtration as part of the conventional treatment train. The samples were taken continuously for 30 hours as shown in Figure A.4. For *Cryptosporidium*

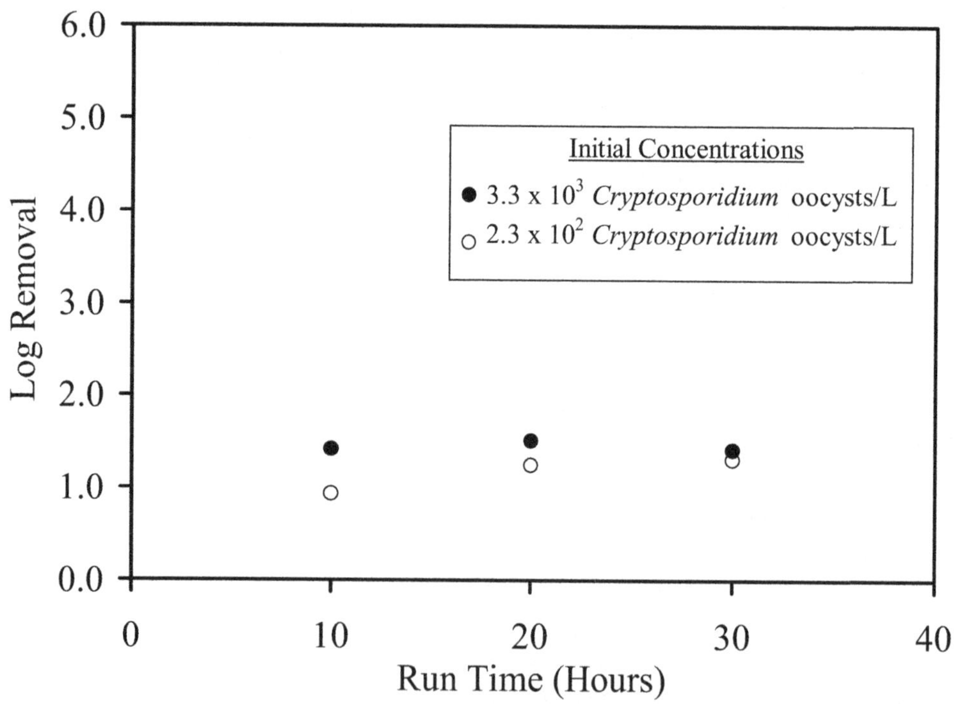

Figure A.7 ***Cryptosporidium*** **oocyst removal through filtration (based on filter effluent grab samples).**

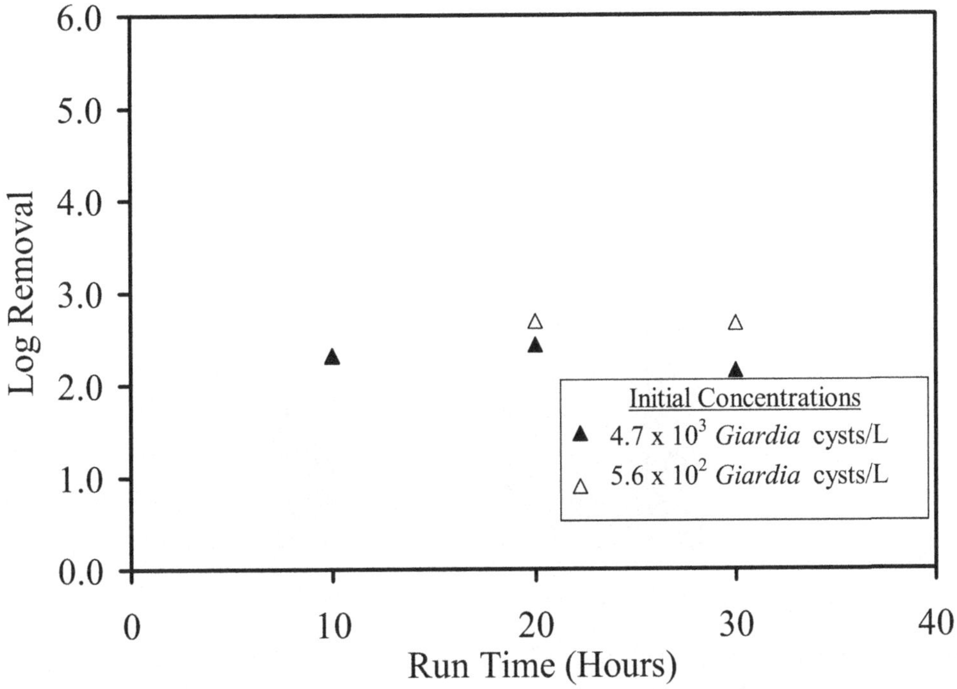

Figure A.8 ***Giardia*** **cyst removal through filtration (based on filter effluent grab samples).**

oocysts, log removals of 2.0 and 1.5 were achieved across filtration with settled water concentrations of 3.3 x 10^3 oocysts/L and 2.3 x 10^2 oocysts/L, respectively. Filtration was able to remove 3.0 log of *Giardia* cysts in both treatment trains.

Comparison of Grab Sampling with Continuous Sampling

Log removals based on grab sampling were compared with those based on continuous sampling (see Figures A.9 and A.10). The observed *Cryptosporidium* oocyst removals based on grab sampling at low initial pathogen concentrations were about the same as the observed removal based on continuous sampling. However, for the high initial pathogen concentration, the removal based on continuous sampling was about 0.5-log higher than the removals based on grab sampling. The same was observed for *Giardia* cysts.

Cryptosporidium **Oocyst Removal versus** ***Giardia*** **Cyst Removal**

The results obtained from the November 2004 experiment were used to generate Figures A.11 to A.15. Figure A.11 compares log removal of *Cryptosporidium* oocysts with log removal of *Giardia* cysts through sedimentation. A paired *t*-test demonstrated that the mean log *Giardia* cyst removal was not significantly different from the mean log *Cryptosporidium* oocyst removal at a 95% confidence level (p-value = 0.10). Figures A.12 and A.13 compare log removal of *Cryptosporidium* oocysts with log removal of *Giardia* cysts through conventional treatment. A paired *t*-test showed that the mean log *Giardia* cyst removal was significantly different from the mean log *Cryptosporidium* oocyst removal at a 95% confidence level (p-value < 0.01). The

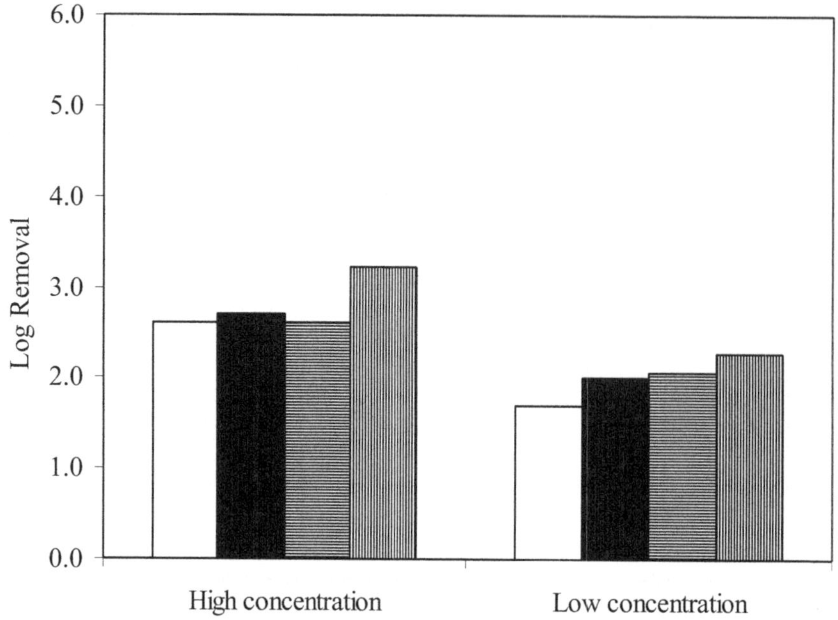

Figure A.9 Comparison of sampling methods for *Cryptosporidium* oocyst removal through conventional treatment. Symbols: □ Grab Sampling at 10 hours, ■ Grab Sampling at 20 hours, ▤ Grab Sampling at 30 hours, ▥ Continuous Sampling from 5 to 35 hours.

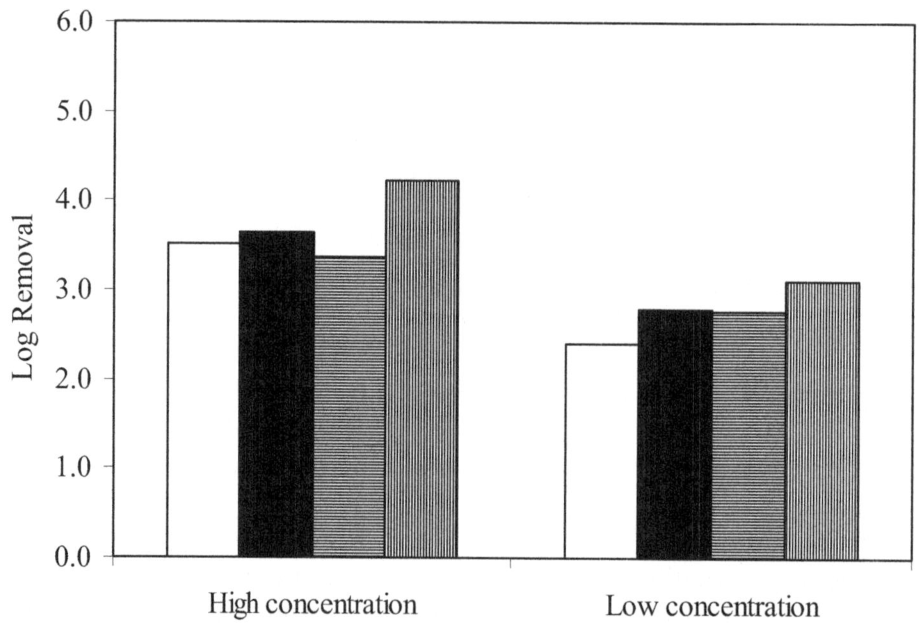

Figure A.10 Comparison of sampling methods for *Giardia* cyst removal through conventional treatment. Symbols: □ Grab Sampling at 10 hours, ■ Grab Sampling at 20 hours, ▤ Grab Sampling at 30 hours, ▥ Continuous Sampling from 5 to 35 hours.

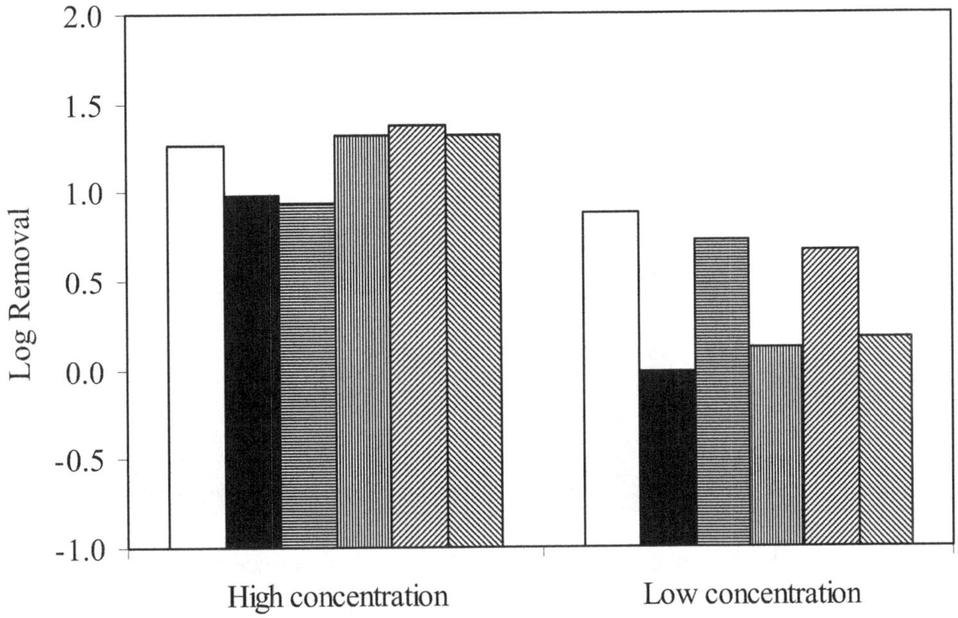

Figure A.11 Comparison of *Cryptosporidium* oocyst and *Giardia* cyst removals through sedimentation. Symbols: □ *Cryptosporidium* at 10 hours, ■ *Giardia* at 10 hours, ▤ *Cryptosporidium* at 20 hours, ▥ *Giardia* at 20 hours, ▨ *Cryptosporidium* at 30 hours, ▧ *Giardia* at 30 hours.

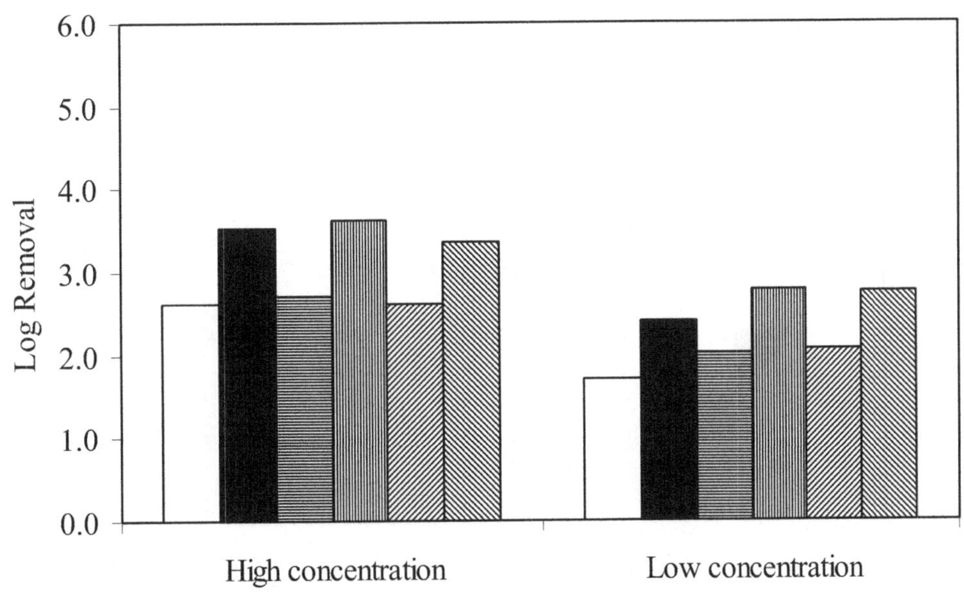

Figure A.12 Comparison of *Cryptosporidium* oocyst and *Giardia* cyst removals through conventional treatment, based on grab samples. Symbols: □ *Cryptosporidium* at 10 hours, ■ *Giardia* at 10 hours, ▤ *Cryptosporidium* at 20 hours, ▥ *Giardia* at 20 hours, ▨ *Cryptosporidium* at 30 hours, ▧ *Giardia* at 30 hours.

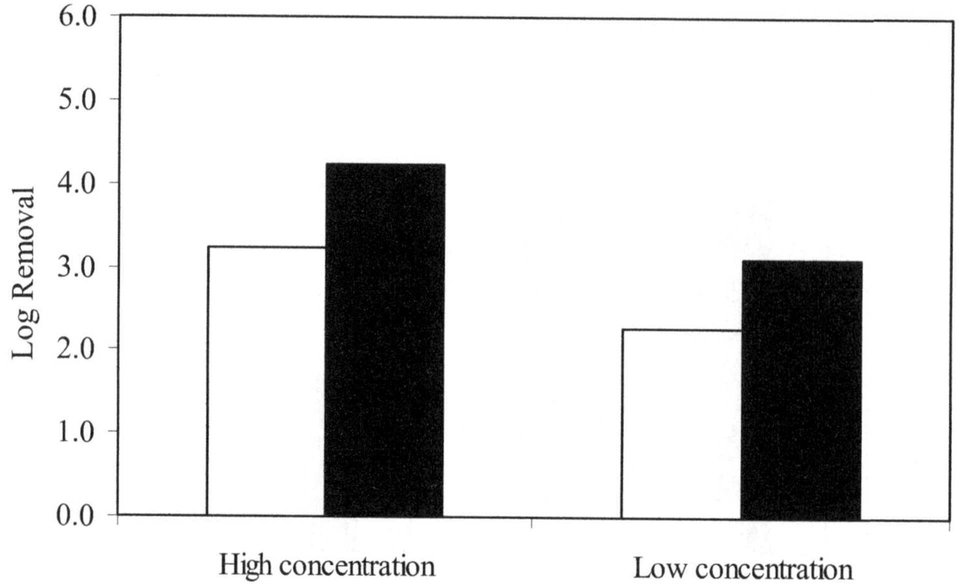

Figure A.13 Comparison of *Cryptosporidium* oocyst and *Giardia* cyst removals through conventional treatment, based on continuous samples. Symbols: □ *Cryptosporidium*, ■ *Giardia*.

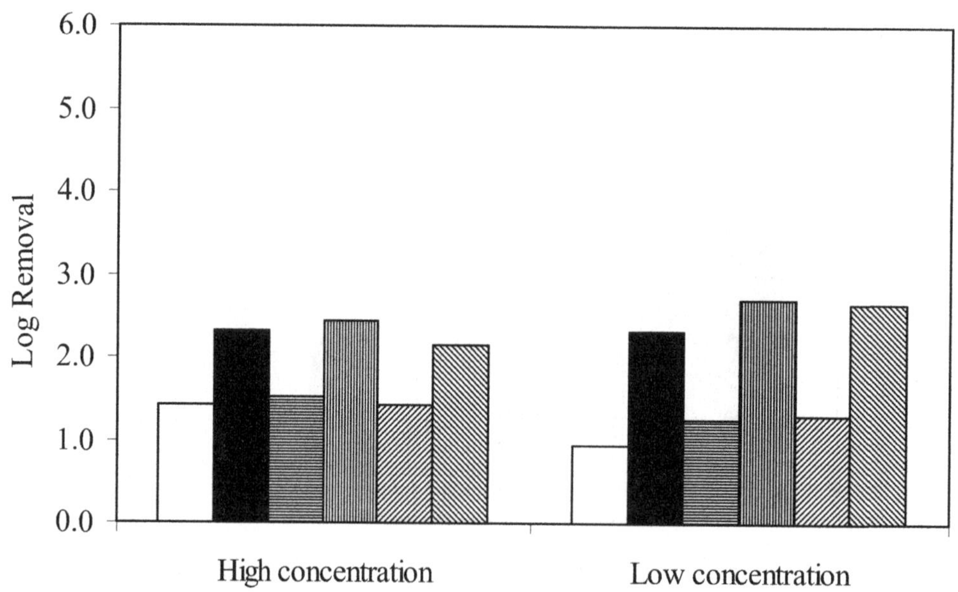

Figure A.14 Comparison of *Cryptosporidium* oocyst and *Giardia* cyst removals through filtration, based on grab samples. Symbols: □ *Cryptosporidium* at 10 hours, ■ *Giardia* at 10 hours, ▤ *Cryptosporidium* at 20 hours, ▥ *Giardia* at 20 hours, ▨ *Cryptosporidium* at 30 hours, ▧ *Giardia* at 30 hours.

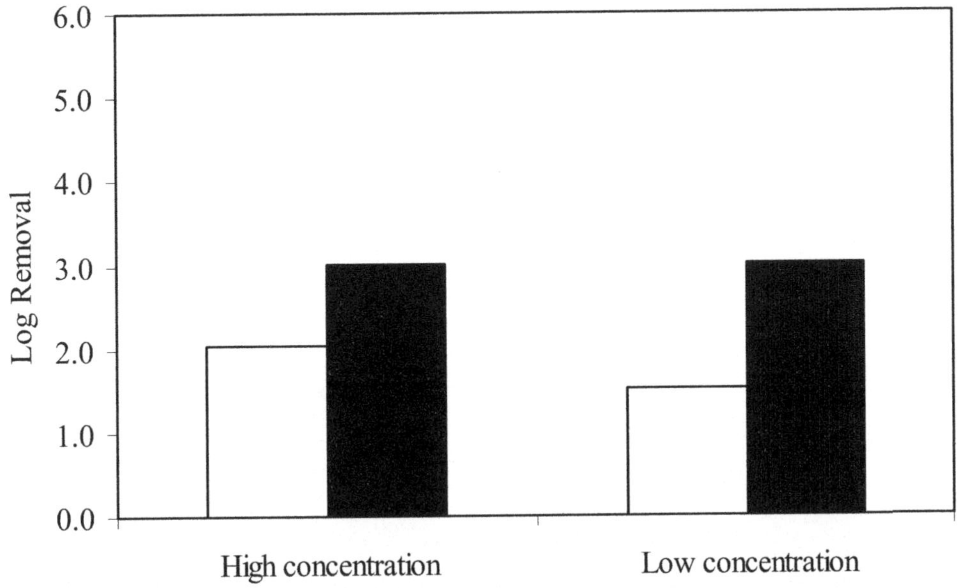

Figure A.15 Comparison of *Cryptosporidium* oocyst and *Giardia* cyst removals through filtration, based on continuous samples. Symbols: □ *Cryptosporidium*, ■ *Giardia*.

mean log *Giardia* cyst removal was 0.8 log higher than the mean log *Cryptosporidium* oocyst removal. Comparison between log removals of *Cryptosporidium* oocysts with log removals of *Giardia* cysts through filtration was shown in Figure A.14 and A.15. A paired *t*-test suggested that *Giardia* cysts were more readily removed than *Cryptosporidium* oocysts through filtration (p-value < 0.01). The mean log *Giardia* cyst removal was 1.2 log higher than the mean log *Cryptosporidium* oocyst removal.

APPENDIX B
EXPERIMENTAL CONDITIONS AND RESULTS FOR THE JANUARY 2005 EXPERIMENT

An alum dose of 50 mg/L was used for the treatment train that had spiked raw water concentrations of 5.3 x 10^2 ± 1.0 x 10^2 *Cryptosporidium* oocysts/L and 3.9 x 10^2 ± 0.5 x 10^2 *Giardia* cysts/L. The other treatment train had an alum dose of 60 mg/L and was spiked with a higher pathogen dose: 7.0 x 10^4 ± 0.8 x 10^4 *Cryptosporidium* oocysts/L and 6.1 x 10^4 ± 1.2 x 10^4 *Giardia* cysts/L. The alum dose was based on jar tests where the source water was supplemented with test pathogens. This was done to ensure that the selected pilot-scale coagulant dose accounted for the concentration of the pathogen feed. Therefore, both treatment trains were run with optimal coagulation conditions. Raw water temperature during the experiment was 6.7 ± 0.5°C. Spiked raw water turbidity averaged 1.59 ± 0.03 ntu for Train 1 and 1.60 ± 0.06 ntu for Train 2 during the experiment.

PATHOGEN REMOVAL FROM ALUM ADDITION THROUGH SEDIMENTATION

Turbidity Results

Figure B.1 shows that sedimentation effluent turbidity in both treatment trains was stable with an average of 1.0 ± 0.1 ntu throughout the entire sampling period. This result is comparable to those observed in previous pilot-scale treatment studies (Edzwald *et al.* 2000, Harrington *et al.* 2001, Dugan *et al.* 2001). A paired *t*-test showed that the sedimentation effluent turbidity in Train 1 was not significantly different from the sedimentation effluent turbidity in Train 2 (p-value = 0.37). These results show that the treatment trains were achieving the same settled water turbidity. In addition, there was no statistically significant trend between settled water turbidity and experimental run time, suggesting that the treatment processes from alum addition through sedimentation were operated at steady-state conditions throughout the experiment.

Pathogen Removals through Sedimentation from Grab Samples

Figures B.2 and B.3 illustrate the effect of spike dose on pathogen removal from alum addition through sedimentation. With an initial *Cryptosporidium* oocyst concentration of 7.0 x 10^4 oocysts/L, an average log removal of 1.1 ± a standard deviation of 0.1 was achieved by the sequence of coagulation, flocculation, and sedimentation. The other treatment train, which ran at the same time with the same treatment configuration but with a lower initial *Cryptosporidium* oocyst concentration of 5.3 x 10^2 oocysts/L, achieved a lower log removal of 0.2 ± 0.1 (also see Figure B.2). The same pattern was observed for *Giardia* cyst removal (see Figure B.3). Treatment trains spiked with *Giardia* cysts at 6.1 x 10^4 cysts/L and 3.9 x 10^2 cysts/L were able to achieve an average log removal of 1.3 ± 0.1 and 0.3 ± 0.0, respectively.

Based on results from a paired *t*-test, the difference between the log removals from the two treatment trains was statistically significant for both microbes (p-value < 0.05). For both organisms, removal was one log higher at the higher spike dose than at the lower spike dose.

The results suggest that pathogen removal was higher in the treatment train receiving the higher pathogen spike dose. One can question whether the observed differences in removal were

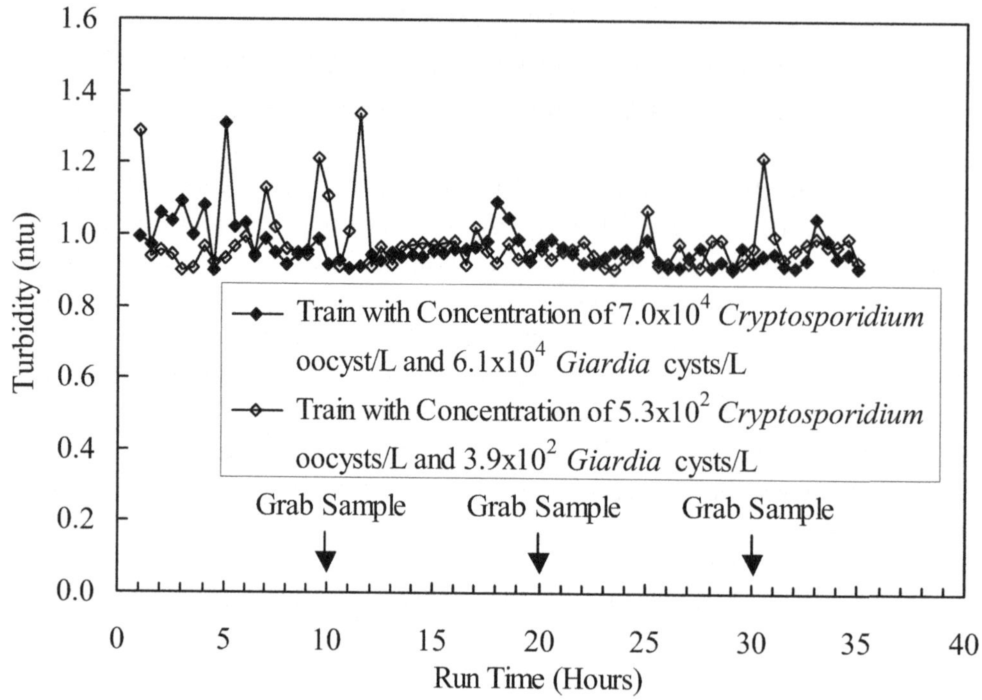

Figure B.1 Turbidity in sedimentation effluent versus run time in the January 2005 experiment. Vertical arrows show times at which grab samples were collected during the experiment.

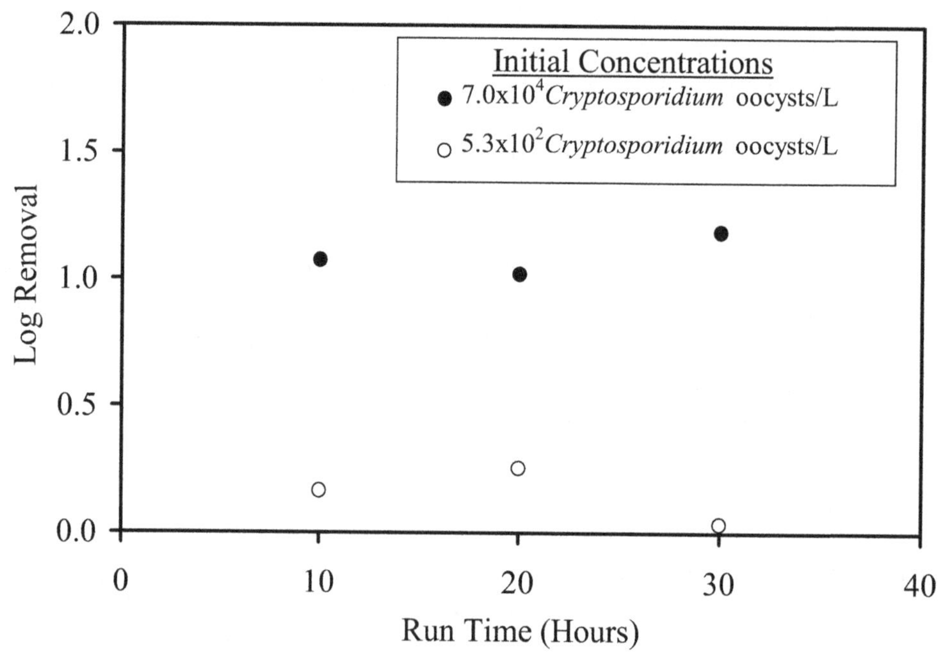

Figure B.2 ***Cryptosporidium*** **oocyst removal through sedimentation based on grab sampling.**

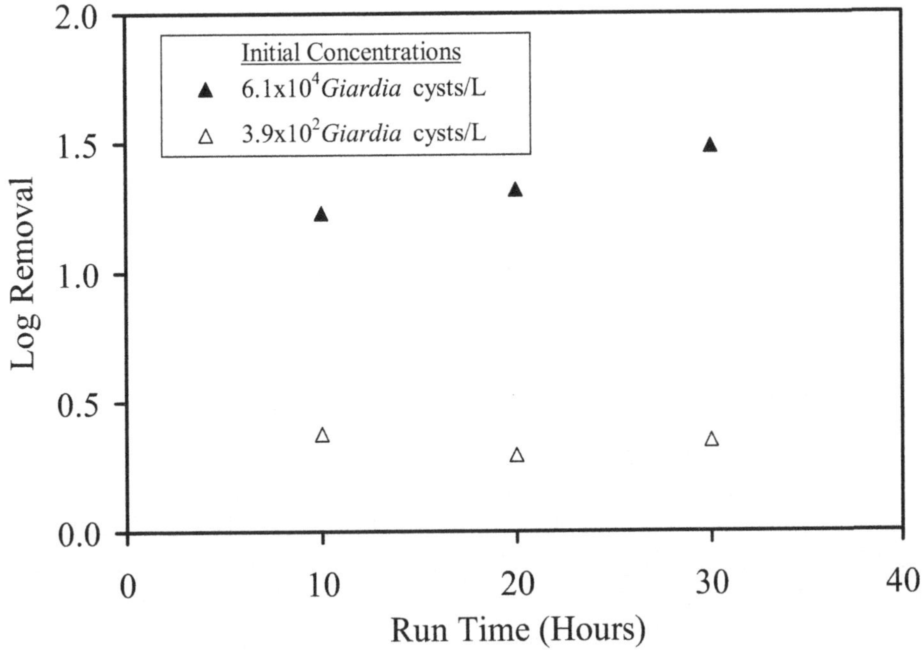

Figure B.3 ***Giardia*** **cyst removal through sedimentation based on grab sampling.**

due to the difference in spike dose or due to the fact that one treatment train might perform better than another treatment train. To address this, multiple regression was performed on data from all experiments to assess whether train-to-train variability was a confounding variable in our studies. The result of this regression analysis suggested that there was no significant difference in performance between treatment train 1 and treatment train 2 at a 95 % confidence level. This result was consistent with previous studies on the pilot plant, which showed that train-to-train experimental error was ± 0.3 log and that this error was random (i.e., no single treatment train consistently performed better than the other train, Harrington *et al.* 2001).

Linear regression was performed to determine if *Cryptosporidium* oocyst and *Giardia* cyst removals were independent of experimental run time. For example, in the treatment train with an initial concentration of 7.0 x 10^4 oocysts/L, log removal of *Cryptosporidium* oocysts and experimental run time were linearly correlated with a slope of 0.006 and a p-value of 0.53. Because the p-value for the estimate of the slope was greater than 0.05, there was no statistically significant relationship between log removal and experimental run time. The same result was obtained for all experiments. This suggests that the pilot plant was operating at near steady-state conditions for those processes upstream of filtration. These steady pathogen removals were consistent with the steady turbidity removals that were presented in the previous section.

PATHOGEN REMOVAL FROM ALUM ADDITION THROUGH GRANULAR MEDIA FILTRATION

Turbidity Results

Figure B.4 shows filter effluent turbidity measurements from the two on-line turbidimeters located in the pilot plant. As noted earlier, one of the on-line turbidimeters was located on the treatment train with the higher pathogen dose while the other on-line turbidimeter was located on the treatment train with the lower pathogen dose. The turbidity trend was generally consistent with the Ginn *et al.* (1992) model described earlier, with ripening and mature operation both being evident on Figure B.4.

During stable operation, filter effluent turbidity levels in both trains were below 0.1 ntu. The average filter effluent turbidity was 0.08 ± 0.02 ntu for the train with the higher pathogen dose and 0.05 ± 0.01 ntu for the train with the lower pathogen dose. The higher pathogen dose was associated with a higher alum dose, which means there were also more aluminum hydroxide particles created in this train. A paired *t*-test was run to evaluate the difference in turbidity between the two treatment trains. The turbidity from the two treatment trains was estimated to be statistically different at more than a 95% confidence level (p-value = 0.00).

To help interpret the pathogen results presented later, Figure B.4 also shows the times at which samples were collected for pathogen analysis during this experiment. As noted earlier, sample collection was initiated after 5 hours of filter run time or when the filter effluent turbidity

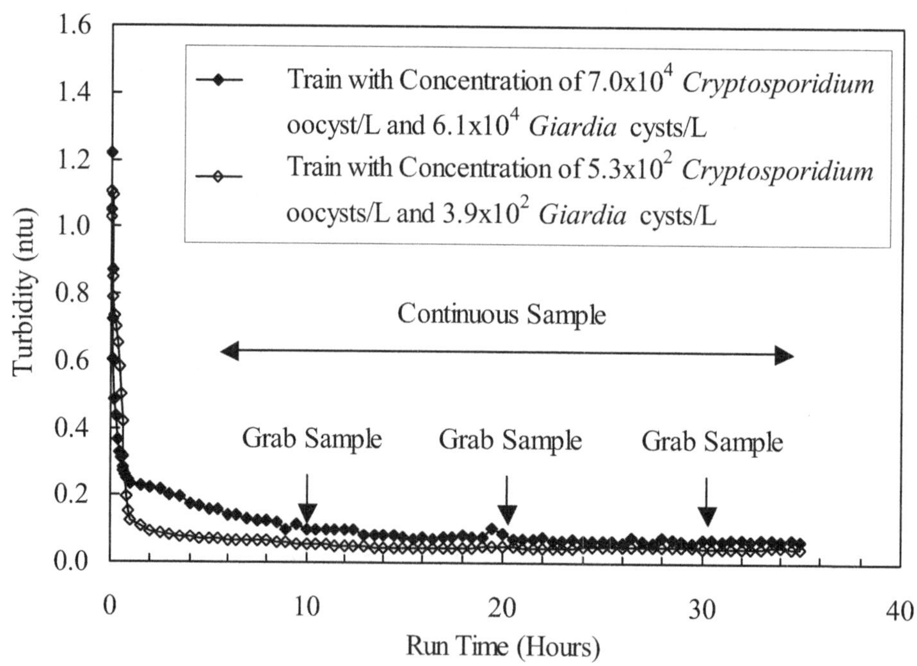

Figure B.4 Turbidity in filtration effluent versus run time in the January 2005 experiment. Vertical arrows show times at which grab samples were collected during the experiment. Horizontal arrows indicate the period of time when centrifuges were operating to collect continuous samples.

dropped below 0.3 ntu in both treatment trains, whichever occurred last. Also, the experiment was halted at 35 hours of run time because previous studies with this pilot plant have shown that turbidity breakthrough or terminal head loss generally occurs between 35 and 40 hours of run time (Harrington *et al.* 2001).

Particle Count Results

Figure B.5 shows filter effluent particle counts for 3 to 7 μm sized particles, which represents the typical size range of *Cryptosporidium* oocysts. As shown in Figure B.5, the treatment train that received the higher pathogen dose had an average particle count of 3.4×10^1 particles/mL, which was about one order of magnitude higher than the average particle count of 4.6×10^0 particles/mL in the treatment train that received the lower pathogen dose. The difference in particle counts between the two treatment trains was significant at a 95 percent confidence level, based on a paired *t*-test. Similar results were observed for particles in the 10 to 15 μm size range, as shown in Figure B.6. This size range represents typical sizes of *Giardia* cysts. As noted earlier, the higher pathogen dose was also associated with a higher alum dose, which means there were more aluminum hydroxide particles created in this train. As with Figure B.4, Figures B.5 and B.6 show the times at which pathogen samples were collected during this experiment.

The observed particle count data over the entire period of the filter run can be used to assess filter performance over time. As seen in Figures B.5 and B.6, there were spike events observed by the particle counters. These spikes are unpredictable and routinely observed in real-world treatment plants. This observation has important implications for grab sampling versus continuous sampling. As noted earlier, grab sampling is useful for observing this particular

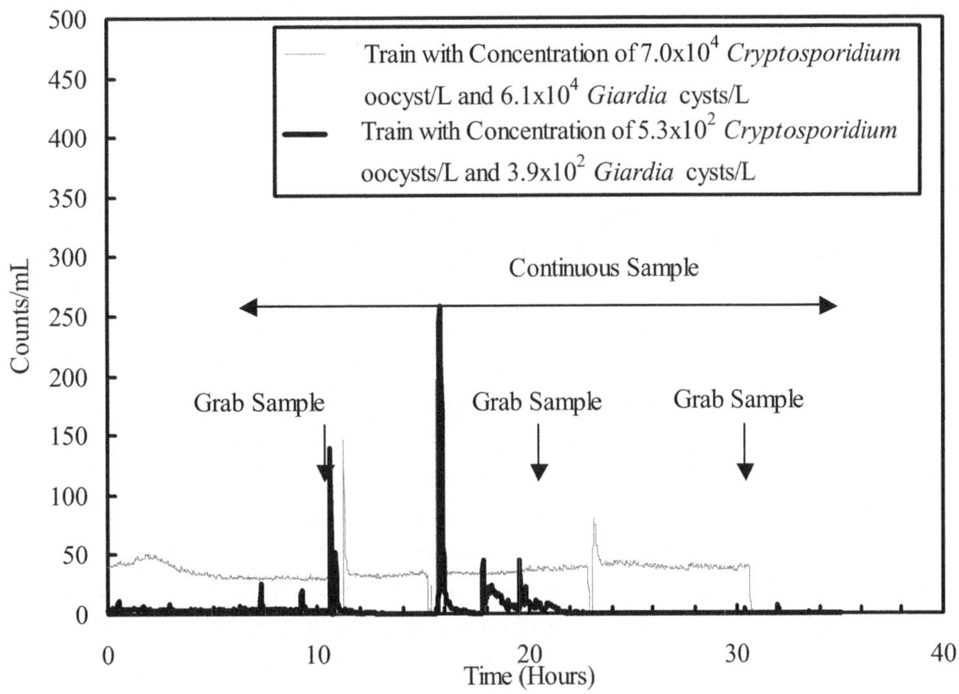

Figure B.5 Counts of 3 to 7 μm size particles in filter effluent (January 2005 experiment).

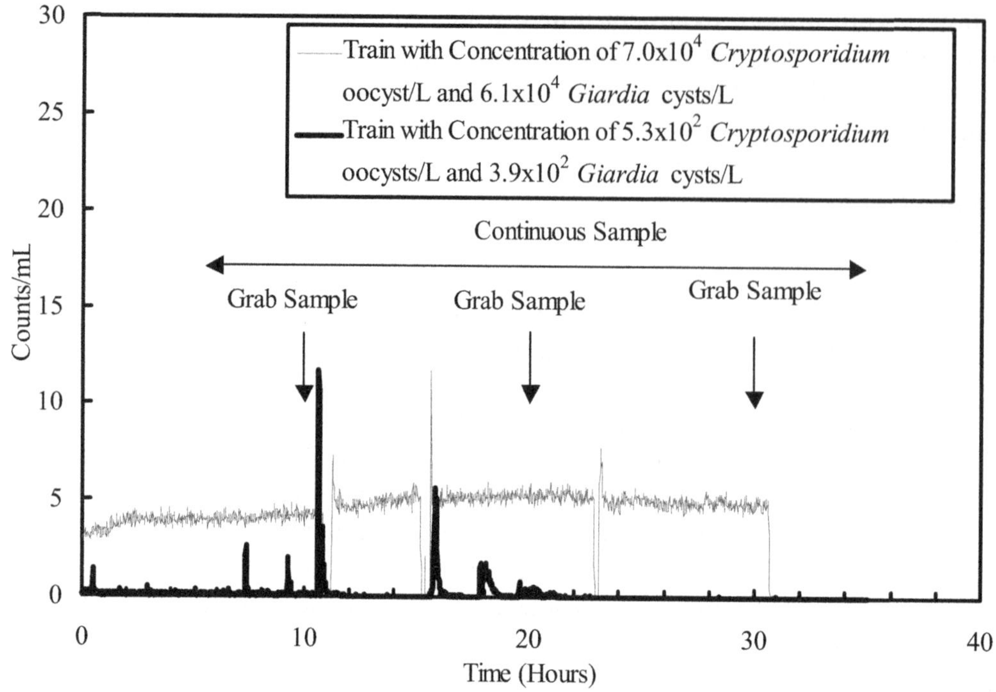

Figure B.6 Counts of 10 to 15 μm size particles in filter effluent (January 2005 experiment).

instant if collected at the proper time while continuous sampling allows for an assessment of average filter performance over the filter run. Therefore, it is likely that grab sampling will not include the spike events shown in these two figures while continuous sampling will include them in the assessment of average performance.

Pathogen Removal through Conventional Treatment from Filter Effluent Grab Samples

Log removal results from grab samples are shown in Figures B.7 and B.8 for spiked Lake Mendota water that went through conventional treatment (*i.e.*, the entire treatment train from alum addition through flocculation, sedimentation, and filtration). Average *Cryptosporidium* oocyst log removals of 3.2 ± 0.5 and 1.1 ± 0.3 were achieved in the treatment trains with initial *Cryptosporidium* oocyst concentrations of 7.0 x 10^4 oocysts/L and 5.3 x 10^2 oocysts/L, respectively (see Figure B.7). In terms of average *Giardia* cyst removal, conventional treatment was able to remove 3.8 ± 0.1 log and 1.5 ± 0.4 log when the respective spike doses were 6.1 x 10^4 cysts/L and 3.9 x 10^2 cysts/L (see Figure B.8). Removal was about two orders of magnitude lower at lower spike doses than at higher spike doses. A paired *t*-test showed that the mean log removals in the train with the higher pathogen concentrations were significantly different at a 95% confidence level from the mean log removals in the train with the lower pathogen concentrations. The continuous sampling results were consistent with the grab sampling trends at these low spike doses of 10^1/L. These results are provided later in this section.

A similar pattern occurred in the November 2004 experiment (see Appendix A). For the March 2005, March 2006, December 2006 and March 2007 experiments, pathogens were not

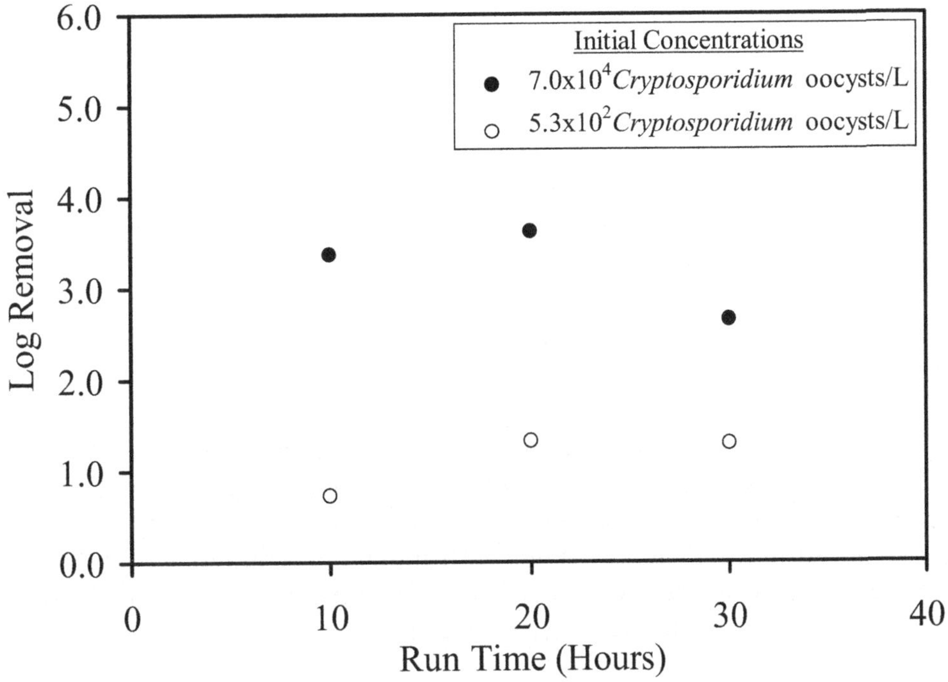

Figure B.7 ***Cryptosporidium*** **oocyst removals through conventional treatment (based on filter effluent grab samples).**

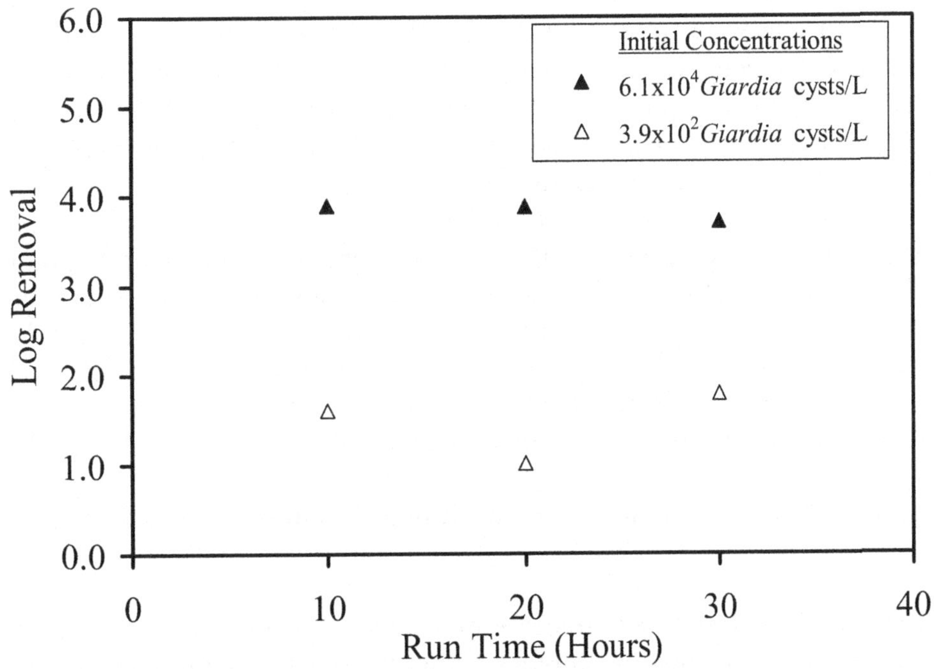

Figure B.8 ***Giardia*** **cyst removal through conventional treatment (based on filter effluent grab samples).**

detected in the grab samples collected from the treatment train with spike doses of about 10^1/L (see Appendices C, D, E and F). However, it is important to note that continuous sampling was able to concentrate pathogens up to the level of detection for most of the samples from that treatment train.

Linear regression was conducted and results showed no correlation between removal and filter run time over the time period studied in any of the experiments. This verifies that the samples were collected during a period of stable filter operation and were consistent with the turbidity and particle count trends shown in Figures B.7 through B.8.

Pathogen Removal through Conventional Treatment for Continuous Samples Collected from Filter Effluent

Log removal results from continuous samples were obtained from spiked Lake Mendota water that went through conventional treatment. The samples were taken continuously for 30 hours as shown in Figures B.4 through B.6. For *Cryptosporidium* oocysts, log removals of 2.6 and 2.2 were achieved in the treatment trains with initial concentrations of 7.0 x 10^4 oocysts/L and 5.3 x 10^2 oocysts/L, respectively. Conventional treatment was able to remove 4.3 log and 3.1 log of *Giardia* cysts when the respective spike doses were 6.1 x 10^4 cysts/L and 3.9 x 10^2 cysts/L.

PATHOGEN REMOVAL FROM SEDIMENTATION THROUGH GRANULAR MEDIA FILTRATION

Pathogen Removal through Filtration from Filter Effluent Grab Samples

Log removal results from grab samples are shown in Figures B.9 and B.10 for spiked Lake Mendota water that went from sedimentation through filtration as part of the conventional treatment. In terms of average *Cryptosporidium* oocyst removal, filtration achieved 2.1 ± 0.5 log and 1.0 ± 0.3 log when the respective pathogen concentrations at the sedimentation tank effluent were 5.7 x 10^3 oocysts/L and 4.1 x 10^2 oocysts/L (see Figure B.9). Average *Giardia* cyst log removals of 2.5 ± 0.1 and 1.1 ± 0.4 were obtained in the treatment trains when *Giardia* cyst concentrations at the effluent of the sedimentation basins were 2.8 x 10^3 cysts/L and 1.8 x 10^2 cysts/L, respectively (see Figure B.10). Removal was about one order of magnitude lower at lower spike doses than at higher spike doses. A paired *t*-test showed that the log removal in the train with the higher oocyst concentration was not significantly different at a 95% confidence level from the log removals in the train with the lower oocyst concentration (p-value = 0.10). However, the same analysis indicated that there was a statistically significant difference in log cyst removals between the two treatment trains (p-value = 0.04).

Pathogen Removal through Filtration for Continuous Samples Collected from Filter Effluent

Log removal results from continuous samples were obtained from spiked Lake Mendota water that went from sedimentation through filtration as part of the conventional treatment process. The samples were taken continuously for 30 hours as shown in Figures B.4 through B.6. For *Cryptosporidium* oocysts, log removals of 1.5 and 2.0 were achieved across filtration

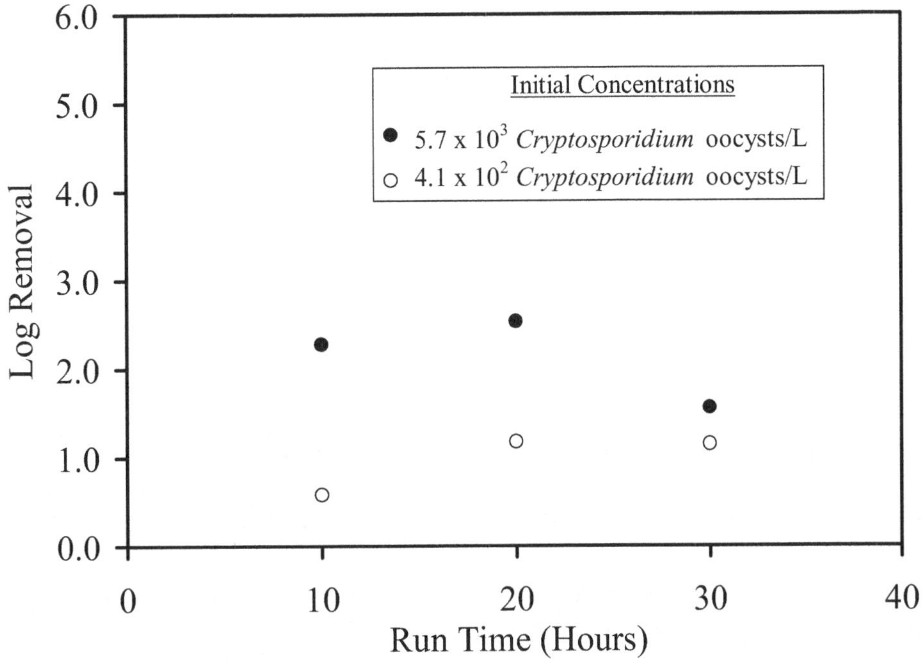

Figure B.9 ***Cryptosporidium*** **oocyst removal through filtration (based on filter effluent grab samples).**

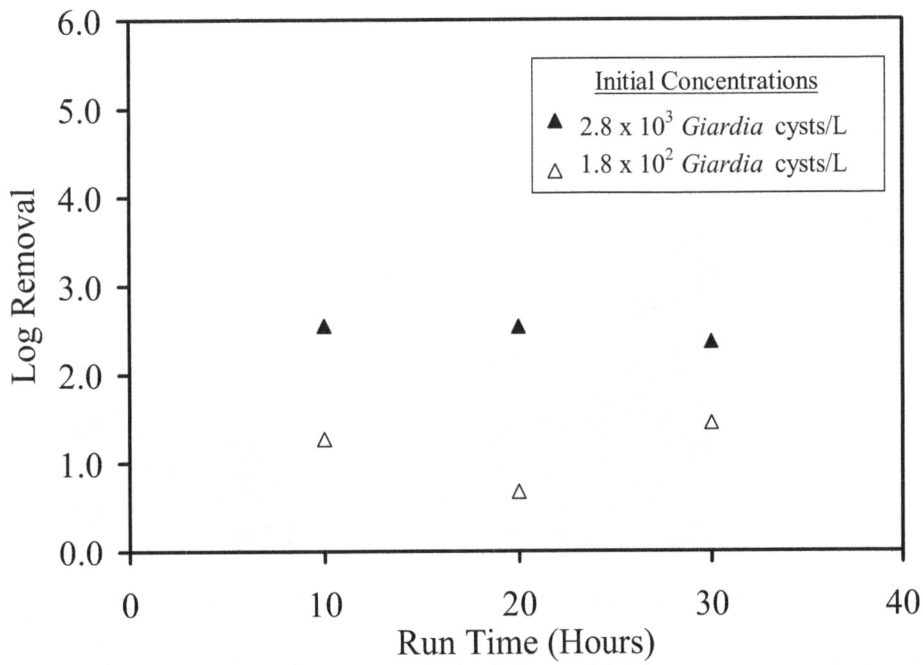

Figure B.10 ***Giardia*** **cyst removal through filtration (based on filter effluent grab samples).**

with sedimentation effluent concentrations of 5.7 x 10^3 oocysts/L and 4.1 x 10^2 oocysts/L, respectively. Filtration was able to remove 2.9 log and 2.8 log of *Giardia* cysts when the respective cyst concentrations were 2.8 x 10^4 cysts/L and 1.8 x 10^2 cysts/L in the effluent of the sedimentation basins.

COMPARISON OF GRAB SAMPLING WITH CONTINUOUS SAMPLING

Figures B.11 and B.12 show a comparison of log removals based on grab sampling with those based on continuous sampling. The observed *Cryptosporidium* oocyst removals based on grab sampling at high initial concentration of pathogens were about the same as the observed removal based on continuous sampling. The same was observed for *Giardia* cyst removal. However, at low initial concentration of pathogens, the removal based on continuous sampling was about 1 log higher than the removals based on grab sampling.

CRYPTOSPORIDIUM OOCYST REMOVAL VERSUS *GIARDIA* CYST REMOVAL

Figure B.13 compares log removal of *Cryptosporidium* oocysts with log removal of *Giardia* cysts through sedimentation for the January 2005 experiment. In this figure, the mean log *Giardia* cyst removal was 0.2 log higher than the mean log *Cryptosporidium* oocyst removal. A paired *t*-test was run at a 95 percent confidence level to determine whether *Giardia* cyst removals were significantly different from *Cryptosporidium* oocyst removals. This test demonstrated that *Giardia* cyst removals were significantly higher than *Cryptosporidium* oocyst removals in this experiment (p-value < 0.01).

Figures B.14 and B.15 compare log removal of *Cryptosporidium* oocysts with log removal of *Giardia* cysts through the conventional treatment train from alum addition through

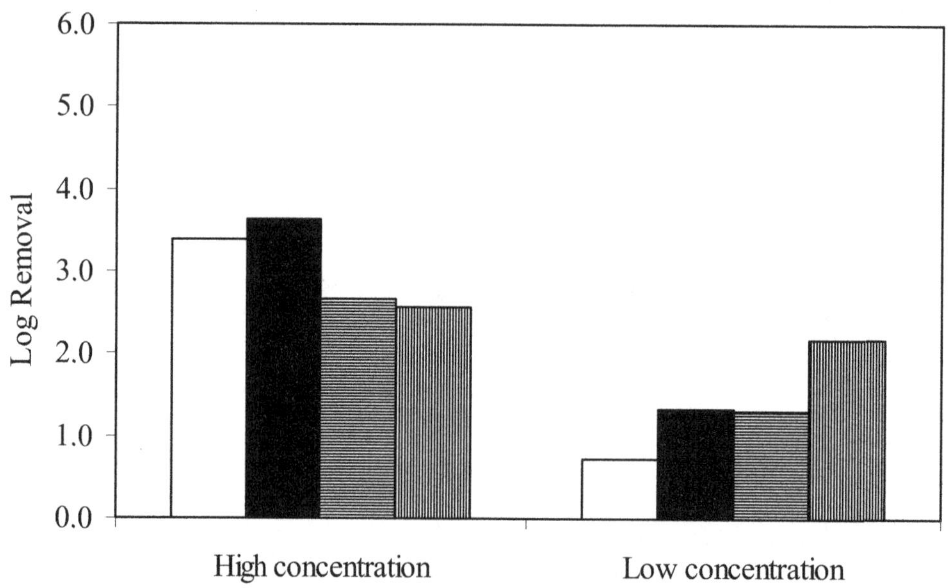

Figure B.11 Comparison of sampling methods for *Cryptosporidium* oocyst removal through conventional treatment. Symbols: □ Grab Sampling at 10 hours, ■ Grab Sampling at 20 hours, ▤ Grab Sampling at 30 hours, ▥ Continuous Sampling from 5 to 35 hours.

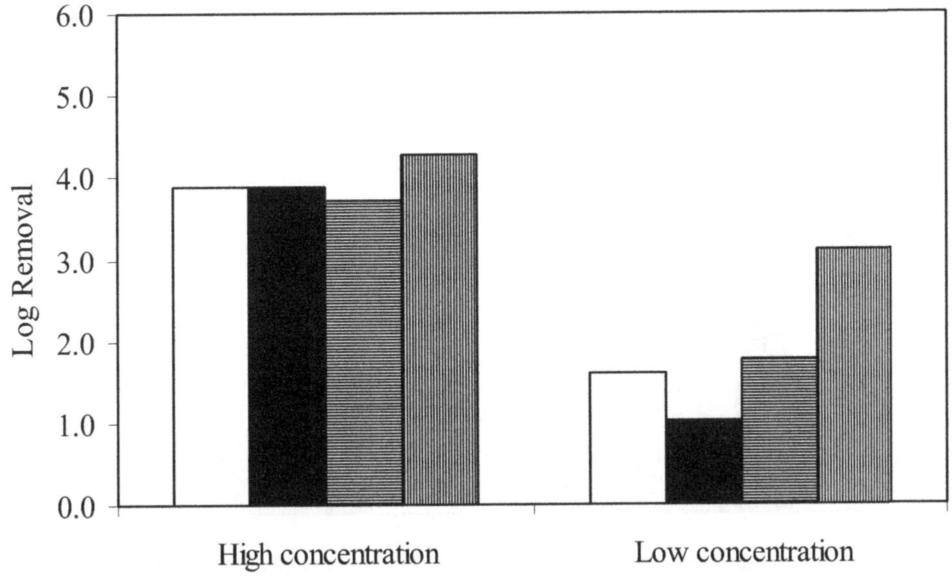

Figure B.12 Comparison of sampling methods for *Giardia* cyst removal through conventional treatment. Symbols: □ Grab Sampling at 10 hours, ■ Grab Sampling at 20 hours, ▤ Grab Sampling at 30 hours, ▥ Continuous Sampling from 5 to 35 hours.

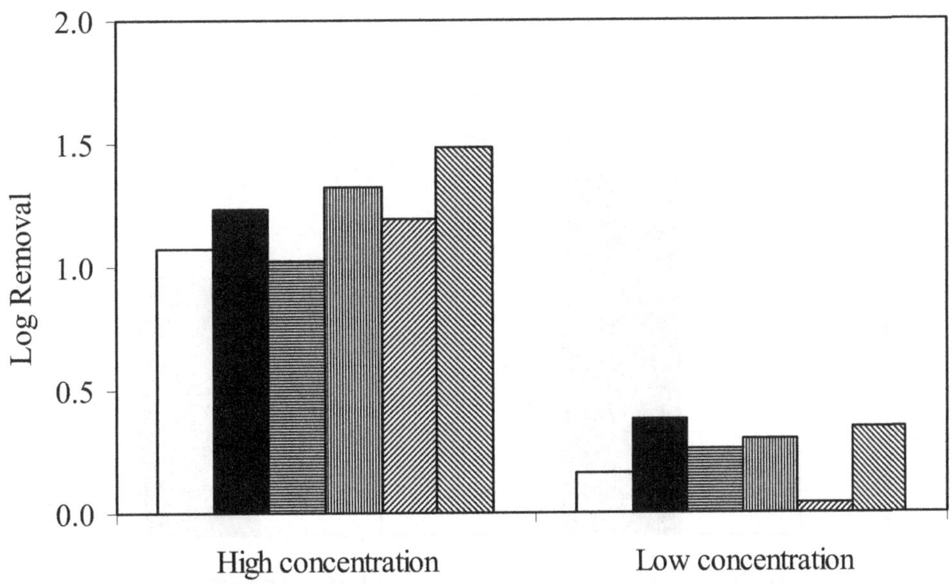

Figure B.13 Comparison of *Cryptosporidium* oocyst and *Giardia* cyst removals through sedimentation. Symbols: □ *Cryptosporidium* at 10 hours, ■ *Giardia* at 10 hours, ▤ *Cryptosporidium* at 20 hours, ▥ *Giardia* at 20 hours, ▨ *Cryptosporidium* at 30 hours, ▧ *Giardia* at 30 hours.

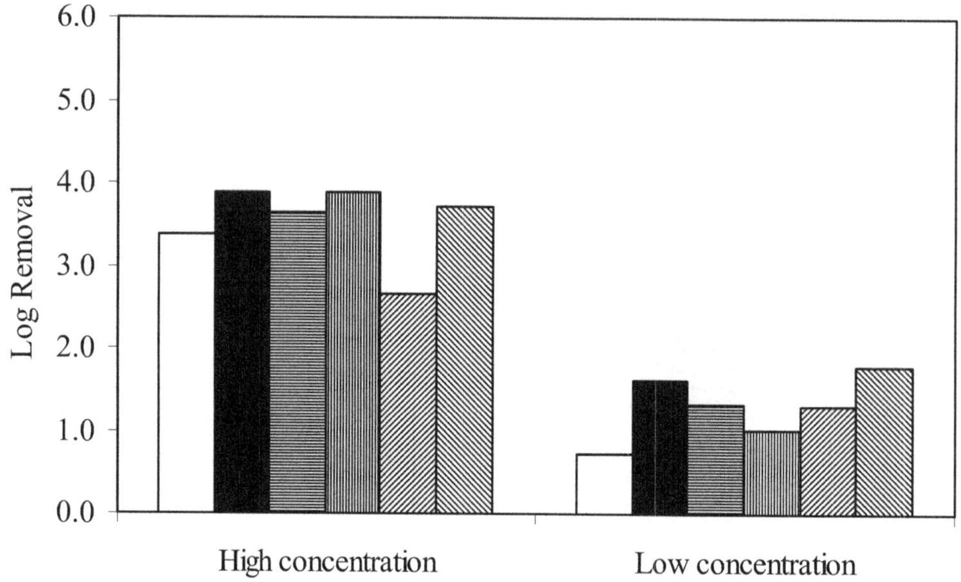

Figure B.14 Comparison of *Cryptosporidium* oocyst and *Giardia* cyst removals through conventional treatment, based on grab samples. Symbols: □ *Cryptosporidium* at 10 hours, ■ *Giardia* at 10 hours, ▤ *Cryptosporidium* at 20 hours, ▥ *Giardia* at 20 hours, ▨ *Cryptosporidium* at 30 hours, ▧ *Giardia* at 30 hours.

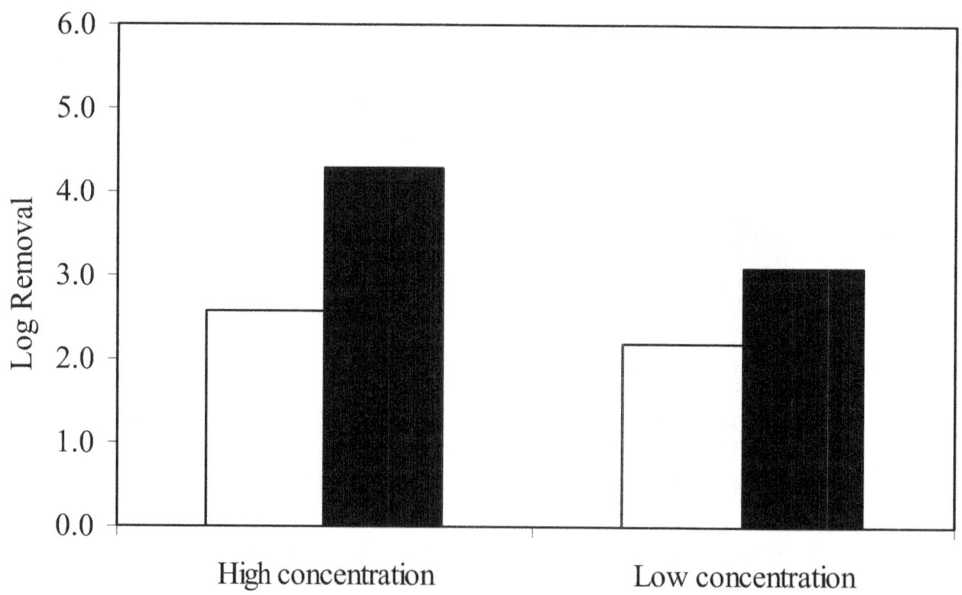

Figure B.15 Comparison of *Cryptosporidium* oocyst and *Giardia* cyst removals through conventional treatment, based on continuous samples. Symbols: □ *Cryptosporidium*, ■ *Giardia*.

filtration. The mean log *Giardia* cyst removal was 0.7-log higher than the mean log *Cryptosporidium* oocyst removal, and this difference was significant at a 95% confidence level (p-value = 0.01). This appears to be consistent with literature reports, where other investigators have generally found *Giardia* cyst removals on the order of 0.0 to 0.2 log higher than removals of *Cryptosporidium* oocysts in pilot scale conventional treatment (Ongerth and Pecoraro 1995, Edzwald *et al.* 2000).

Log removals attained through the filtration process were calculated from pathogen counts in the sedimentation effluent and in the filter effluent. A comparison of oocyst removal with cyst removal through filtration is shown in Figures B.16 and B.17. A paired t-test suggested that the log removal of *Giardia* cysts was not significantly different from the log removal of *Cryptosporidium* oocysts (p-value = 0.06).

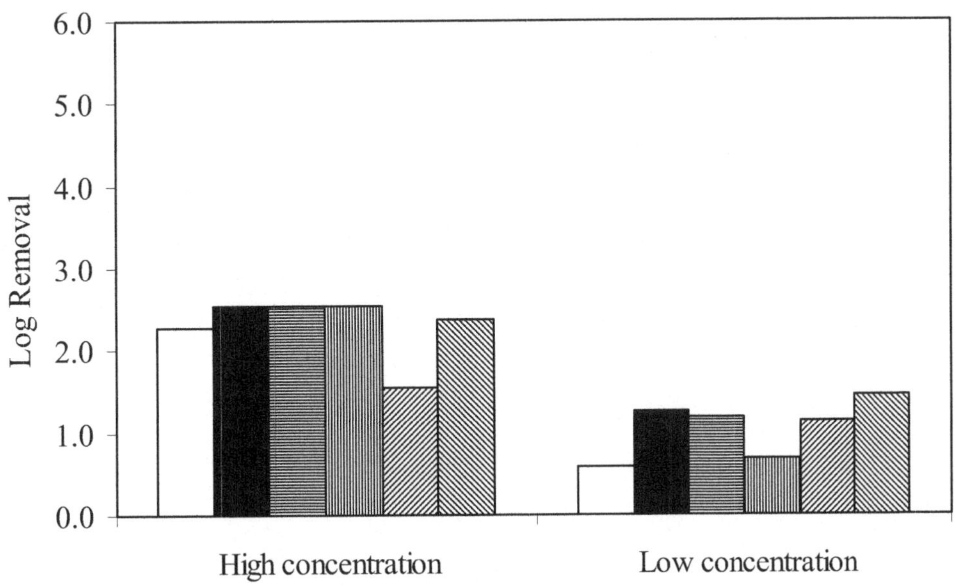

Figure B.16 Comparison of *Cryptosporidium* oocyst and *Giardia* cyst removals through filtration, based on grab samples. Symbols: □ *Cryptosporidium* at 10 hours, ■ *Giardia* at 10 hours, ▤ *Cryptosporidium* at 20 hours, ▥ *Giardia* at 20 hours, ▨ *Cryptosporidium* at 30 hours, ▧ *Giardia* at 30 hours.

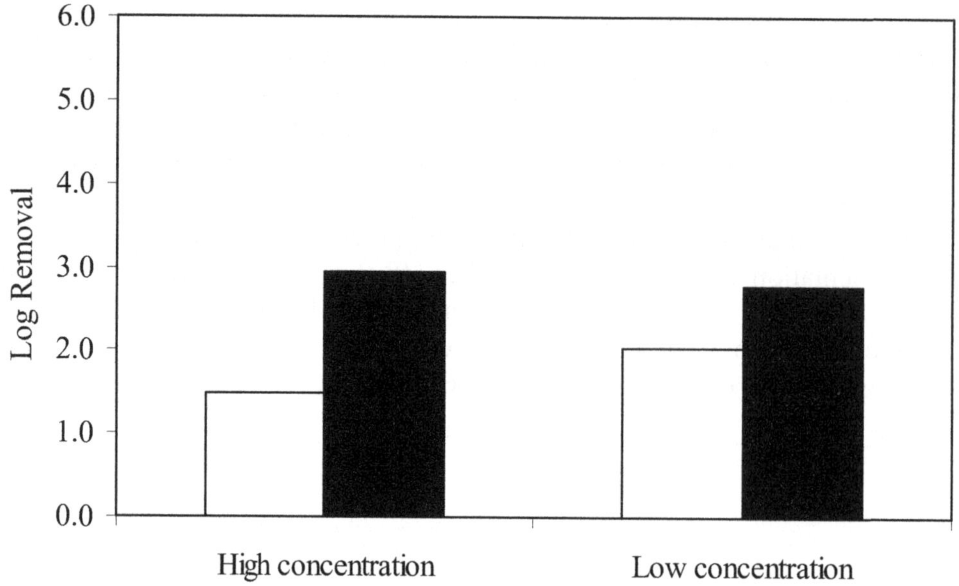

Figure B.17 Comparison of *Cryptosporidium* oocyst and *Giardia* cyst removals through filtration, based on continuous samples. Symbols: □ *Cryptosporidium*, ■ *Giardia*.

APPENDIX C
EXPERIMENTAL CONDITIONS AND RESULTS FOR THE MARCH 2005 EXPERIMENT

The treatment train 1 was spiked with 5.7 x 10^3 ± 1.4 x 10^3 *Cryptosporidium* oocysts/L and 4.4 x 10^3 ± 1.6 x 10^3 *Giardia* cysts/L. The other treatment train (Train 2) was spiked with 1.6 x 10^1 ± 1.7 x 10^1 *Cryptosporidium* oocysts/L and 2.2 x 10^1 ± 2.2 x 10^1 *Giardia* cysts/L. The alum dose was set to 60 mg/L for both treatment trains in this experiment and was based on jar tests where the source water was supplemented with test pathogens. This was done to ensure that the selected pilot-scale coagulant dose accounted for the concentration of the pathogen feed. Therefore, both treatment trains were run with optimal coagulation conditions. Raw water temperature during the experiment was 9.4 ± 0.5°C. Spiked raw water turbidity averaged 0.47 ± 0.03 ntu for Train 1 and 0.47 ± 0.05 ntu for Train 2 during the experiment.

Pathogen Removal from Alum Addition through Sedimentation

Turbidity Results

Turbidity in the sedimentation effluent averaged 0.34 ± 0.04 ntu for Train 1 and 0.36 ± 0.02 ntu for Train 2 (see Figure C.1). These results show that the treatment trains were achieving the same effluent turbidity (*p*-value = 0.115). In addition, the results suggest that the treatment processes from alum addition through sedimentation were operated at steady-state conditions

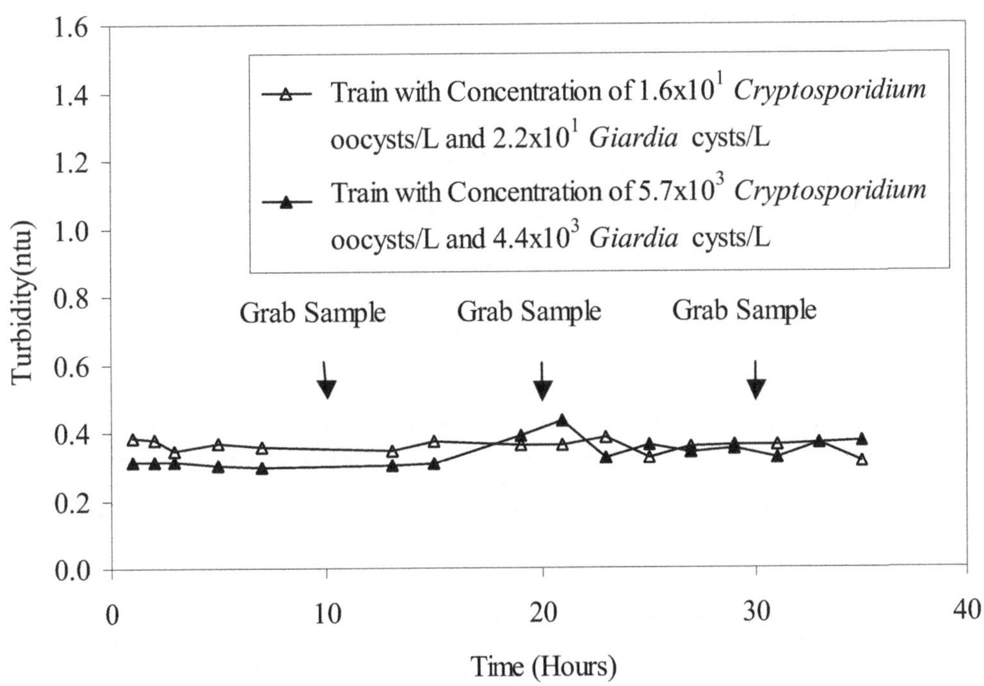

Figure C.1 Turbidity in sedimentation effluent versus run time.

throughout the experiment. Treatment train 1 reduced UV254 by 75 percent from 0.05 cm^{-1} to 0.01 cm^{-1}. For treatment train 2 there was no change in UV254 value.

Pathogen Removals through Sedimentation from Grab Samples

The effect of spike dose on pathogen removal through sedimentation is shown in Table C.1. With initial concentrations of $5.7 \times 10^3 \pm 1.4 \times 10^3$ oocysts/L and $4.4 \times 10^3 \pm 1.6 \times 10^3$ cysts/L, mean log removals of 1.0 ± 0.1 and 1.2 ± 0.1 were obtained for *Cryptosporidium* oocysts and *Giardia* cysts, respectively. For the lower spike doses of $1.6 \times 10^1 \pm 1.7 \times 10^1$ oocysts/L and $2.2 \times 10^1 \pm 2.2 \times 10^1$ cysts/L, no *Cryptosporidium* oocysts or *Giardia* cysts were found in samples taken after sedimentation. Therefore, the oocyst and cyst concentrations were below detection limit, but not necessarily equal to zero. The lowest possible removal would occur if the oocyst and cyst concentrations were just below the detection limit. This implies that the log removal must be greater than the difference between the log of the initial concentration and the log of the detection limit. For example, the average initial log concentration of oocysts in this experiment was $\log_{10}(1.6 \times 10^1) = 1.2$ and the average log of the detection limit was $\log_{10}(2.0 \times 10^0) = 0.3$. Detection limit of 2.0×10^0 was calculated by knowing that our analytical method can detect down to a single pathogen in the sample. For this case, we analyzed 10 mL from a 50 mL sample that had undergone centrifugation from a 10 L grab sample. As noted earlier, the final volume from centrifugation resulted in a 200 mL sample. In this case, our detection limit is 1 oocyst/10 mL of sample, so in 10 L sample we would have 2 oocyst/L. By this calculation, we would calculate that the log removal must have been greater than 0.9. However, it is inconclusive that the train with the lower initial concentration had a lower log removal than the train with the higher initial concentration.

Pathogen Removal from Alum Addition through Granular Media Filtration

Turbidity Results

Filter effluent turbidity data are shown in Figure C.2. The average filter effluent turbidity was 0.04 ± 0.01 ntu for the train with the higher pathogen dose and was 0.16 ± 0.01 ntu for the

Table C.1
Pathogen removal through sedimentation for the March 2005 experiment.

Filter Run Time (hours)	*Cryptosporidium* Log Removal		*Giardia* Log Removal	
	Spike Dose = 5.7×10^3 oocysts/L	Spike Dose = 1.6×10^1 oocysts/L	Spike Dose = 4.4×10^3 cysts/L	Spike Dose = 2.2×10^1 cysts/L
10	1.1	> 0.9	1.3	> 1.1
20	1.0	> 0.9	1.1	> 1.1
30	0.8	> 0.9	1.2	> 1.1

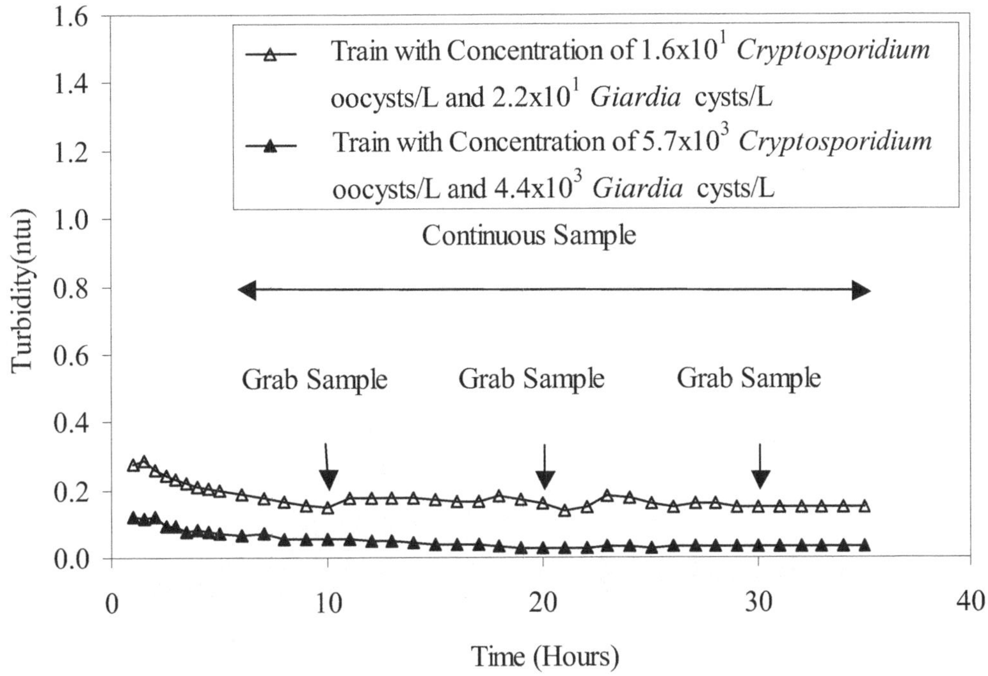

Figure C.2 Turbidity in filtration effluent versus run time in the March 2005 experiment.

train with the lower pathogen dose. The paired *t*-test showed that the mean filter effluent turbidity of the train with the higher pathogen dose was significantly different from the mean filter effluent turbidity of the train with the lower pathogen dose (p-value < 0.01).

Particle Count Results

Filter effluent particle count data are shown in Figures C.3 and C.4 for 3 to 7 μm sized particles and 10 to 15 μm sized particles, respectively. The results included only 7 hours of data because of a problem encountered with the connection of the particle counter to the flow stream. However, the results show the performance of the filter over the period of time shown in the figure. As shown in Figure C.3, the treatment train that received the higher pathogen dose had higher particle counts in the 3 to 7 μm size range than the treatment train that received the lower pathogen dose. Similar results were observed for particles in the 10 to 15 μm size range. The results are consistent with the other experiments.

Pathogen Removal through Conventional Treatment, Based on Filter Effluent Grab Samples

Pathogen removals through coagulation, flocculation, sedimentation and filtration are shown in Table C.2 for grab samples. The train with initial concentrations of 5.7 x 10^3 oocysts/L and 4.4 x 10^3 cysts/L achieved log removals of 2.5 ± 0.1 for *Cryptosporidium* oocysts and 3.0 ± 0.1 for *Giardia* cysts. For the train with 1.6 x 10^1 oocysts/L and 2.2 x 10^1 cysts/L, microscopy did not detect *Cryptosporidium* oocysts or *Giardia* cysts found in grab samples taken from the filter effluent. However, FC sorting-microscopy was able to enumerate cysts and oocysts in

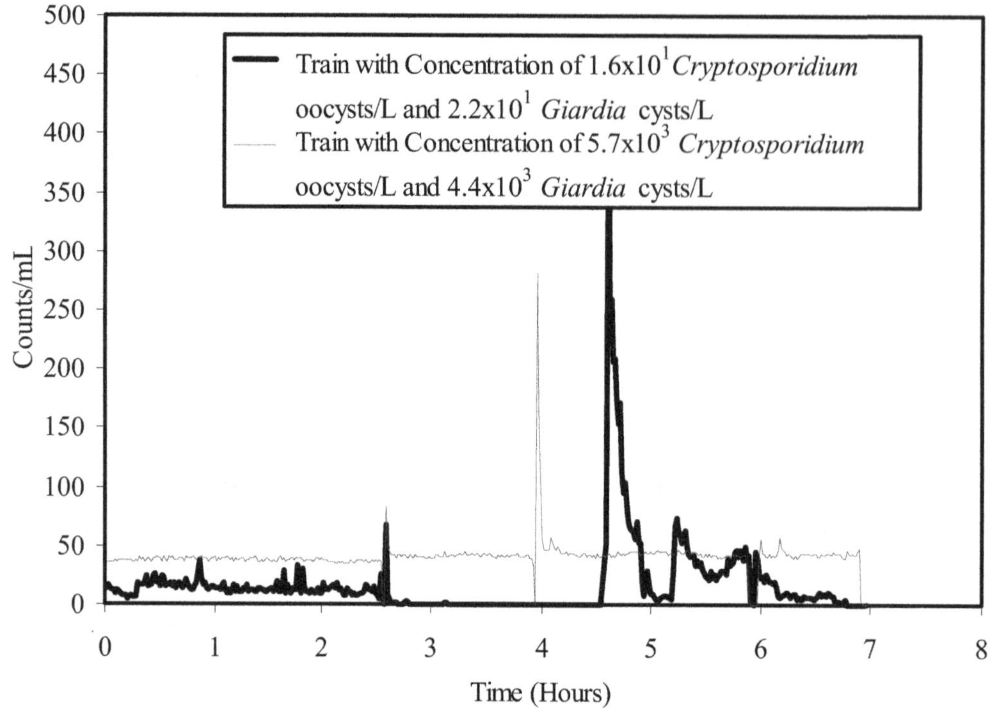

Figure C.3 3 to 7 μm size particles in filtration effluent versus run time in the March 2005 experiment.

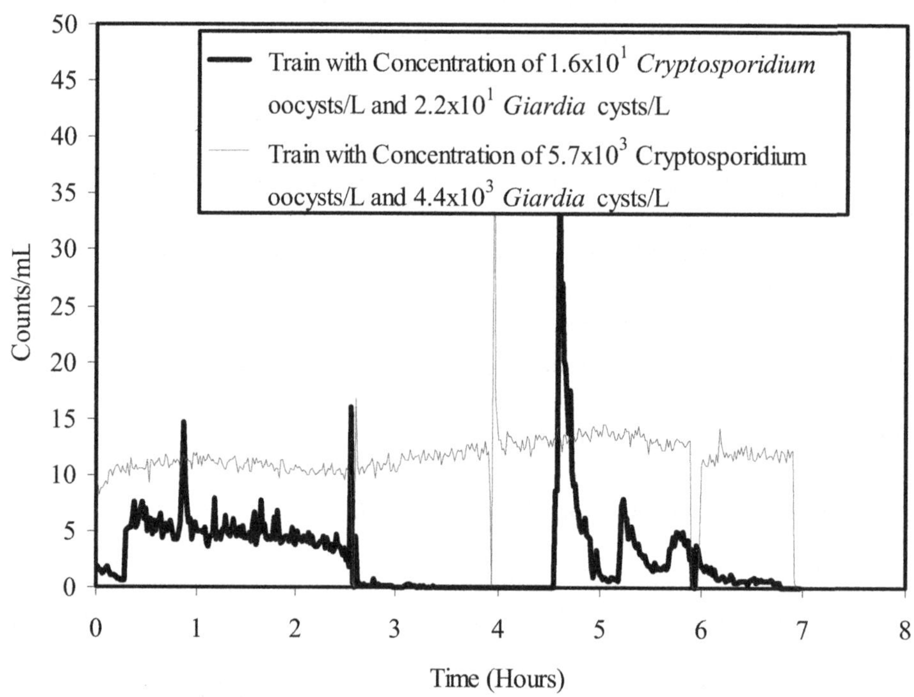

Figure C.4 10 to 15 μm size particles in filtration effluent versus run time in the March 2005 experiment.

Table C.2
Pathogen removal through conventional treatment for the March 2005 experiment.

Filter Run Time (hours)	*Cryptosporidium* Log Removal		*Giardia* Log Removal	
	Spike Dose = 5.7×10^3 oocysts/L	Spike Dose = 1.6×10^1 oocysts/L	Spike Dose = 4.4×10^3 cysts/L	Spike Dose = 2.2×10^1 cysts/L
10	2.4	> 0.3	3.0	> 0.9
20	2.5	> 0.8	2.9	> 1.3
30	2.4	> 0.3	3.1	> 1.7

these grab samples. As mentioned earlier, when samples were below detection limit, log removals were calculated from the mean log number of pathogens in raw water samples and from the average log of the detection limit. In this case, the average log of the detection limit was the result from averaging the detectable number from FC sorting-microscopy and the detection limit from microscopy.

Pathogen Removal through Conventional Treatment for Continuous Samples Collected from Filter Effluent

Log removal results from continuous samples were taken continuously over 30 hours after the ripening phase was complete. The sample times were noted on Figure C.2. Higher removals of *Cryptosporidium* oocysts and *Giardia* cysts were achieved with higher initial spiking doses (5.7×10^3 oocysts/L and 4.4×10^3 cysts/L). For the high concentration train, log removals of 2.6 and 3.2 were observed for *Cryptosporidium* oocysts and *Giardia* cysts, respectively. As expected, removal was lower at the lower spike dose (1.6×10^1 oocysts/L and 2.2×10^1 cysts/L) than at higher spike doses (5.7×10^3 oocysts/L and 4.4×10^3 cysts/L). Conventional treatment was able to remove 1.5 log and 1.9 log of *Cryptosporidium* oocysts and *Giardia* cysts, respectively, when the lower spike doses were seeded.

Pathogen Removal from Sedimentation through Granular Media Filtration

Pathogen Removal through filtration from Filter Effluent Grab Samples

When pathogen concentrations in the sedimentation basin effluent were 5.9×10^2 oocysts/L and 2.6×10^2 cysts/L, filtration achieved average log removals of 1.5 ± 0.1 and 1.8 ± 0.1, respectively. In the lower concentration train, pathogen concentrations in the settled water were below the detection limit and, thus, log removal across filtration could not be calculated.

Pathogen Removal through filtration for Continuous Samples Collected from Filter Effluent

Log removal results from continuous samples were obtained from spiked Lake Mendota water that went from sedimentation to filtration as part of the conventional treatment train. The samples were taken continuously for 30 hours as shown in Figure C.2. For *Cryptosporidium* oocysts, log removals of 1.6 were achieved across filtration with a filter influent concentration of 5.9 x 10^2 oocysts/L. Filtration was able to remove 2.0 log of *Giardia* cysts when cyst concentrations were 2.6 x 10^2 cysts/L.

Comparison of Grab Sampling with Continuous Sampling

Figures C.5 and C.6 show a comparison of log removals based on grab sampling with those based on continuous sampling. The observed *Cryptosporidium* oocyst removals based on grab sampling at high initial concentration of pathogens were about the same as the observed removal based on continuous sampling. For *Giardia* cysts, the log removals based on both types of sampling were also about the same in the high concentration train. At low initial concentration of pathogens, continuous sampling produced detectable concentrations while grab sampling did not.

Cryptosporidium Oocyst Removal versus *Giardia* Cyst Removal

The results obtained from the March 2005 experiment were used to generate Figures C.7 through C.11. Figure C.7 compares log removal of *Cryptosporidium* oocysts with log removal of *Giardia* cysts through sedimentation. A paired *t*-test demonstrated that the mean log *Giardia* cyst removal was not significantly different from the mean log *Cryptosporidium* oocyst removal (*p*-value = 0.08). Figures C.8 and C.9 compare log removal of *Cryptosporidium* oocysts with log removal of *Giardia* cysts through conventional treatment. A paired *t*-test showed that *Giardia* cysts were more readily removed than *Cryptosporidium* oocysts at a 95% confidence level (*p*-

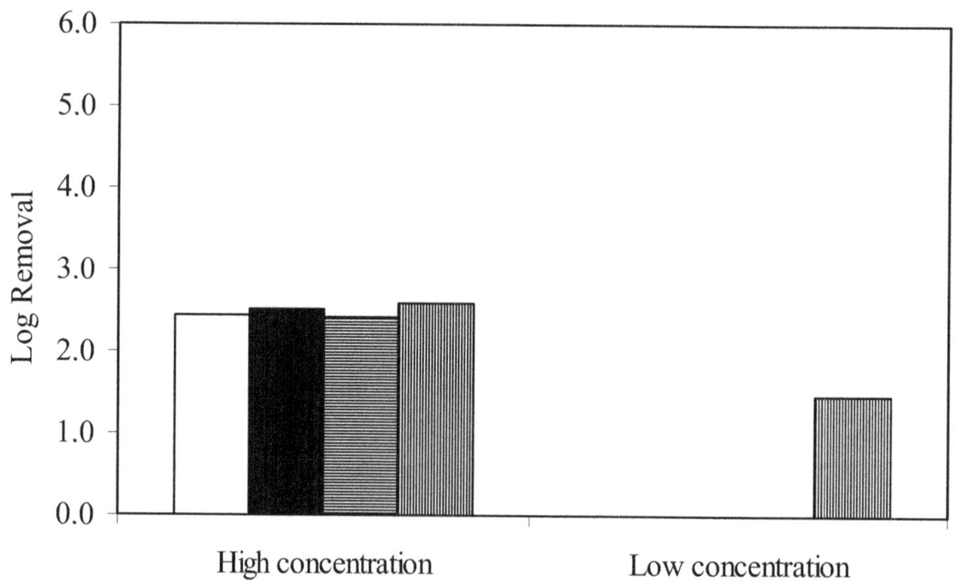

Figure C.5 Comparison of sampling methods for *Cryptosporidium* oocyst removal through conventional treatment. Symbols: □ Grab Sampling at 10 hours, ■ Grab Sampling at 20 hours, ⊟ Grab Sampling at 30 hours, ▥ Continuous Sampling from 5 to 35 hours.

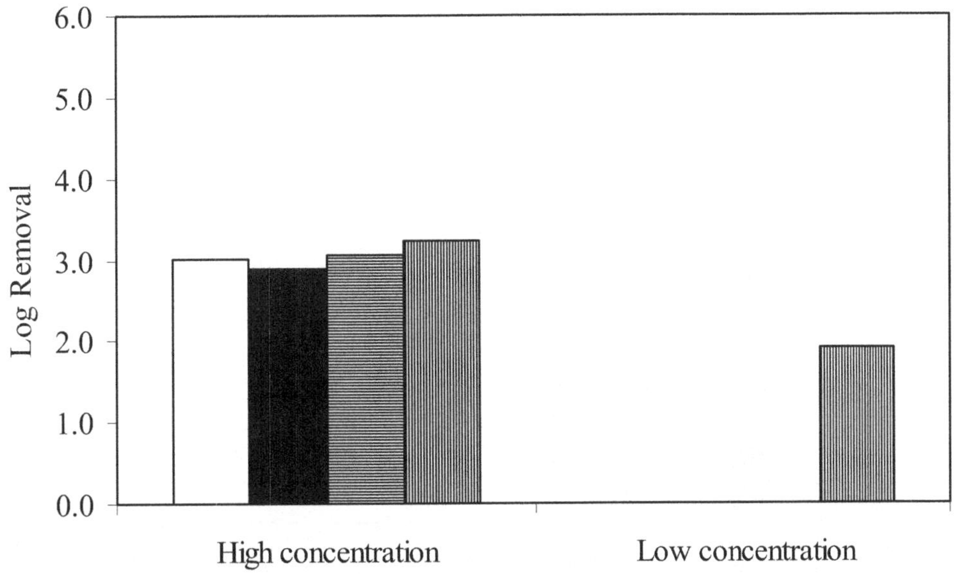

Figure C.6 Comparison of sampling methods for *Giardia* cyst removal through conventional treatment. Symbols: □ Grab Sampling at 10 hours, ■ Grab Sampling at 20 hours, ▤ Grab Sampling at 30 hours, ▥ Continuous Sampling from 5 to 35 hours.

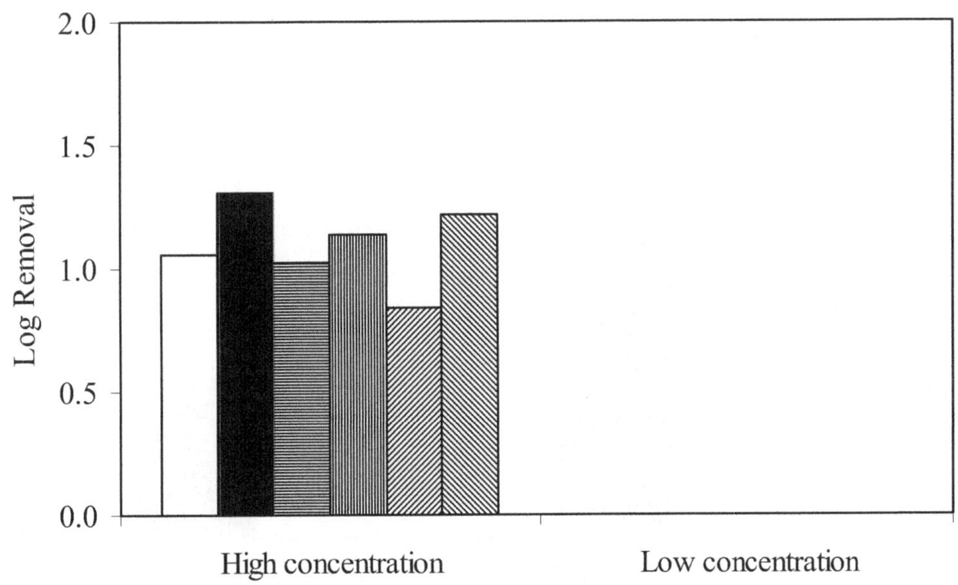

Figure C.7 Comparison of *Cryptosporidium* oocyst and *Giardia* cyst removals through sedimentation. Symbols: □ *Cryptosporidium* at 10 hours, ■ *Giardia* at 10 hours, ▤ *Cryptosporidium* at 20 hours, ▥ *Giardia* at 20 hours, ▨ *Cryptosporidium* at 30 hours, ▧ *Giardia* at 30 hours.

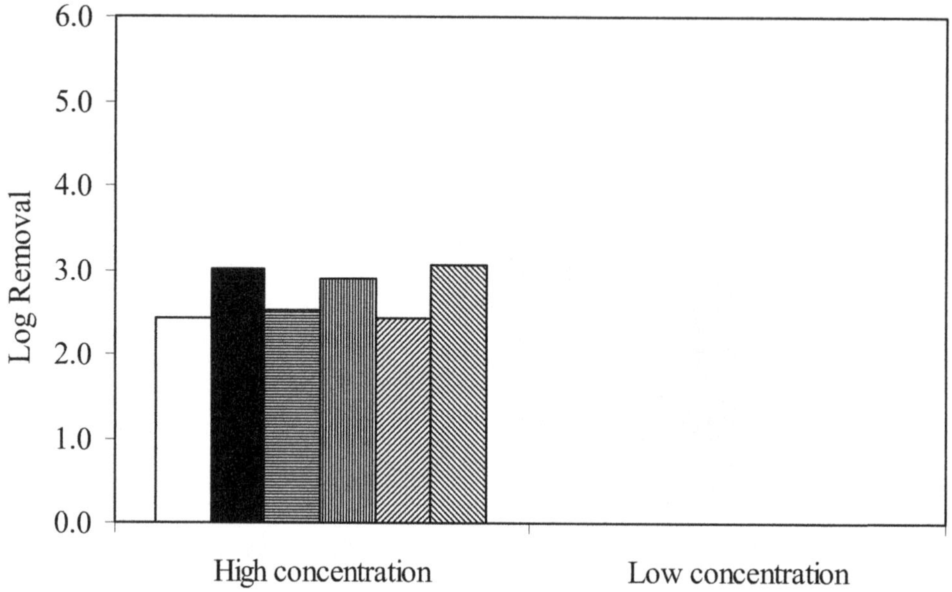

Figure C.8 **Comparison of *Cryptosporidium* oocyst and *Giardia* cyst removals through conventional treatment, based on grab samples. Symbols: □ *Cryptosporidium* at 10 hours, ■ *Giardia* at 10 hours, ▤ *Cryptosporidium* at 20 hours, ▥ *Giardia* at 20 hours, ▨ *Cryptosporidium* at 30 hours, ▧ *Giardia* at 30 hours.**

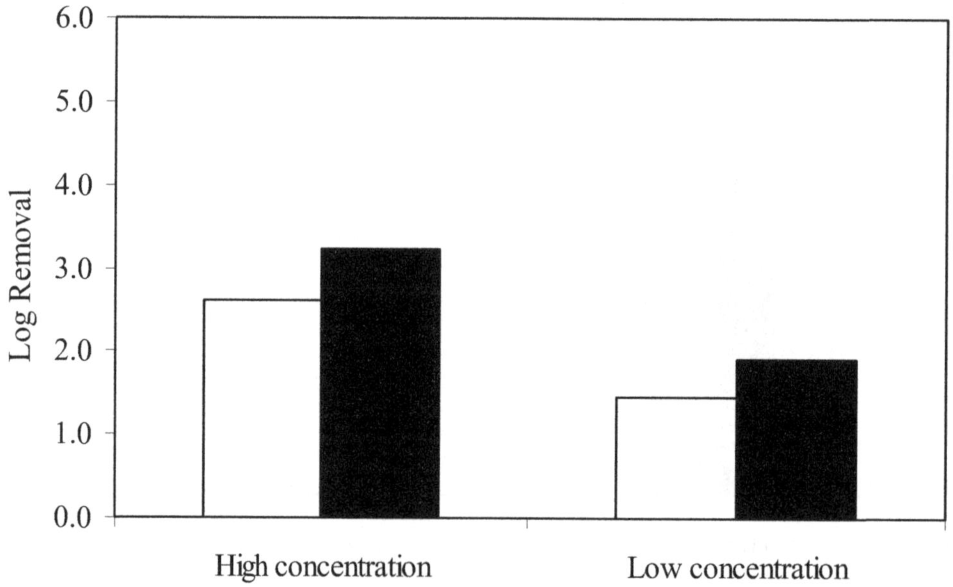

Figure C.9 **Comparison of *Cryptosporidium* oocyst and *Giardia* cyst removals through conventional treatment, based on continuous samples. Symbols: □ *Cryptosporidium*, ■ *Giardia*.**

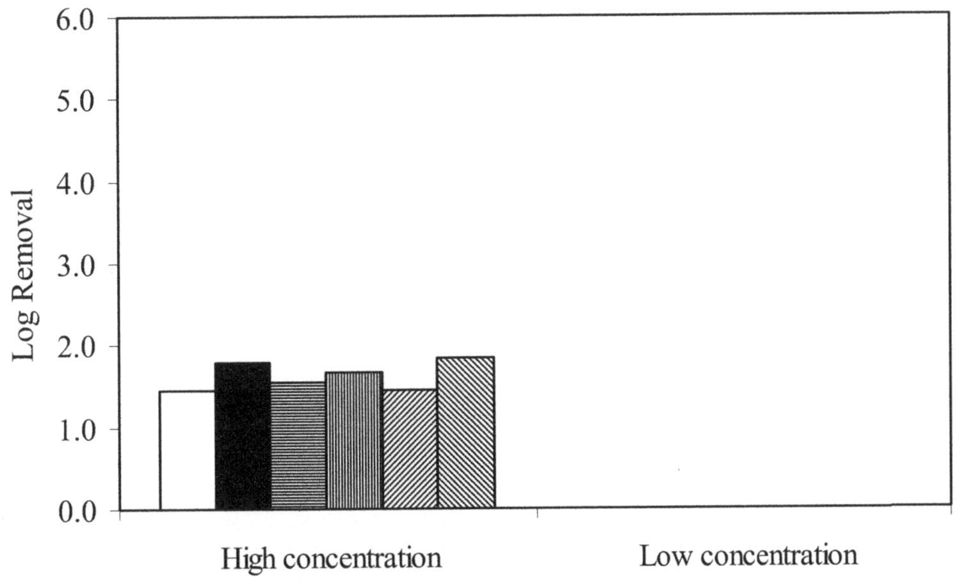

Figure C.10 Comparison of *Cryptosporidium* oocyst and *Giardia* cyst removals through filtration, based on grab samples. Symbols: □ *Cryptosporidium* at 10 hours, ■ *Giardia* at 10 hours, ▤ *Cryptosporidium* at 20 hours, ▥ *Giardia* at 20 hours, ▨ *Cryptosporidium* at 30 hours, ▧ *Giardia* at 30 hours.

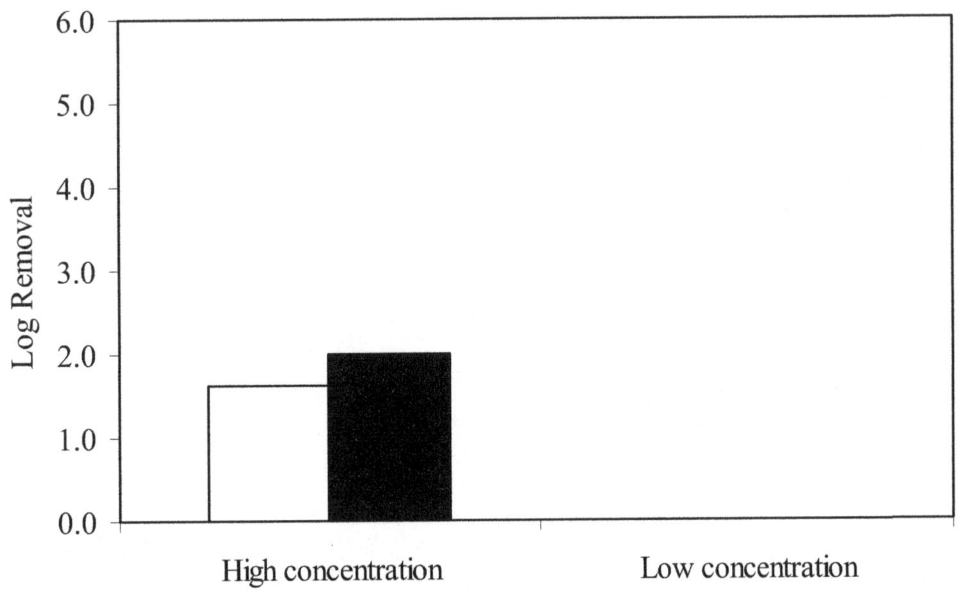

Figure C.11 Comparison of *Cryptosporidium* oocyst and *Giardia* cyst removals through filtration, based on continuous samples. Symbols: □ *Cryptosporidium*, ■ *Giardia*.

value < 0.01). The mean log *Giardia* cyst removal was 0.6 log higher than the mean log *Cryptosporidium* oocyst removal. Figures C.10 and C.11 show the same comparison through filtration. The paired *t*-test again suggested that *Giardia* cysts were more readily removed than *Cryptosporidium* oocysts through filtration (*p*-value = 0.01). In this case, the mean log *Giardia* cyst removal was 0.3 log higher than the mean log *Cryptosporidium* oocyst removal.

APPENDIX D
EXPERIMENTAL CONDITIONS AND RESULTS FOR THE MARCH 2006 EXPERIMENT

The alum dose for the treatment train (Train 1) with initial concentration of $2.3 \times 10^1 \pm 2.5 \times 10^1$ *Cryptosporidium* oocysts/L and $6.4 \times 10^0 \pm 3.2 \times 10^0$ *Giardia* cysts/L was set to 70 mg/L. The other treatment train (Train 2), spiked with $3.0 \times 10^3 \pm 5.6 \times 10^2$ *Cryptosporidium* oocysts/L and $2.2 \times 10^3 \pm 7.2 \times 10^2$ *Giardia* cysts/L, had an alum dose of 80 mg/L. The alum dose was based on jar tests where the source water was supplemented with test pathogens. This was done to ensure that the selected pilot-scale coagulant dose accounted for the concentration of the pathogen feed. Therefore, both treatment trains were run with optimal coagulation conditions. Raw water temperature during the experiment was 7.0 ± 1.2°C. Spiked raw water turbidity averaged 0.61 ± 0.25 ntu for the low concentration train and 0.53 ± 0.06 ntu for the high concentration train during the experiment.

Pathogen Removal from Alum Addition through Sedimentation

Turbidity Results

Turbidity in the sedimentation effluent averaged 0.60 ± 0.21 ntu for the low spike train and 0.49 ± 0.07 ntu for the high spike train (see Figure D.1). These results show that the low spike train had periodic stability problems, likely caused by mixer malfunctions at the rapid mix

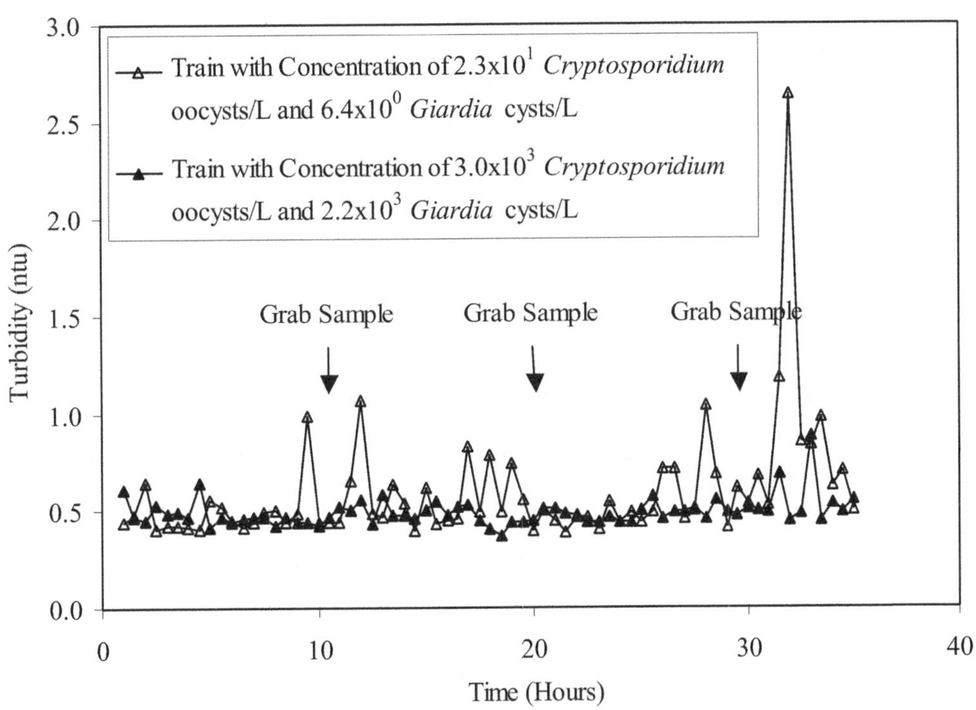

Figure D.1 Turbidity in sedimentation effluent versus run time.

tank. In both treatment trains, treatment reduced UV254 by 33 percent from 0.06 cm^{-1} to 0.04 cm^{-1}.

Pathogen Removal through Sedimentation from Grab Samples

Table D.1 shows the effect of spike dose on pathogen removal through sedimentation. Log removals of 1.3 ± 0.0 and 1.5 ± 0.1 were achieved for oocysts and cysts, respectively, with initial concentrations of 3.0 x 10^3 oocysts/L and 2.2 x 10^3 cysts/L. With initial concentrations of 2.3 x 10^1 oocysts/L and 6.4 x 10^0 cysts/L, there were no *Cryptosporidium* oocysts or *Giardia* cysts found in the settled water. As mentioned in Appendix C, log removals for these samples were calculated from the mean log number of pathogens in raw water samples and the mean log of detection limit.

Pathogen Removal from Alum Addition through Granular Media Filtration

Turbidity Results

Filter effluent turbidity data are shown in Figure D.2. The average filter effluent turbidity was 0.14 ± 0.05 ntu for the train with the higher pathogen dose and was 0.19 ± 0.14 ntu for the train with the lower pathogen dose. The treatment train with the lower pathogen dose had more variation in turbidity values than the treatment train with the higher pathogen dose. This might have resulted from a mixing problem in the rapid mix basin. A paired *t*-test showed that the mean filter effluent turbidity of the train with the higher pathogen dose was significantly different from the mean filter effluent turbidity of the train with lower pathogen dose (p-value = 0.012).

Particle Count Results

Filter effluent particle count data are shown in Figures D.3 and D.4 for 3 to 7 μm sized particles and for 10 to 15 μm sized particles, respectively. As shown in Figure D.3, the treatment train that received the higher pathogen dose had an average particle count of 4.7 x 10^1 particles/mL, which was comparable to the average particle count of 3.9 x 10^1 particles/mL in

Table D.1
Pathogen removal through sedimentation for the March 2006 experiment.

Filter Run Time (hours)	*Cryptosporidium* Log Removal		*Giardia* Log Removal	
	Spike Dose = 3.0 x 10^3 oocysts/L	Spike Dose = 2.3 x 10^1 oocysts/L	Spike Dose = 2.2 x 10^3 cysts/L	Spike Dose = 6.4 x 10^0 cysts/L
10	1.3	> 1.2	1.5	> 0.7
20	1.3	> 1.2	1.6	> 0.7
30	1.3	> 1.2	1.5	> 0.7

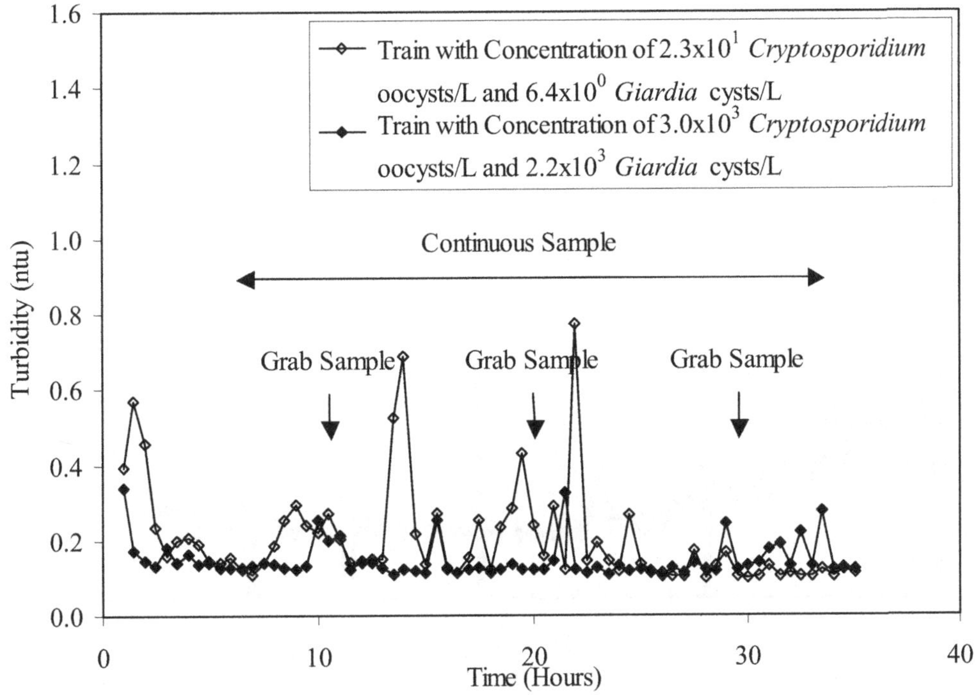

Figure D.2 Turbidity in filtration effluent versus time.

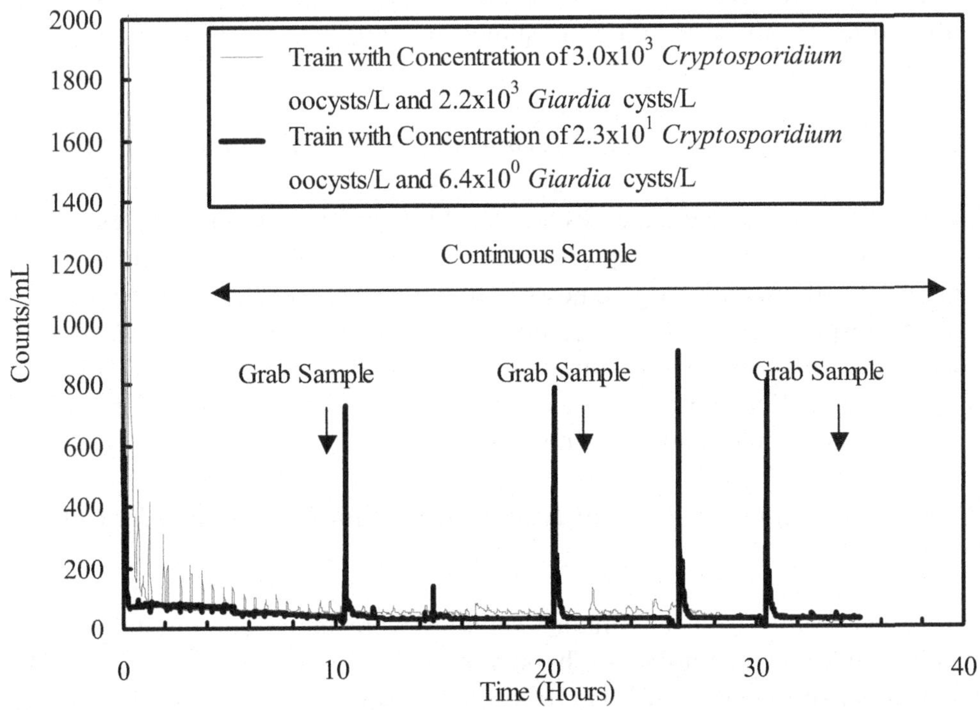

Figure D.3 3 to 7 μm size particles in filtration effluent versus run time in the March 2006 experiment.

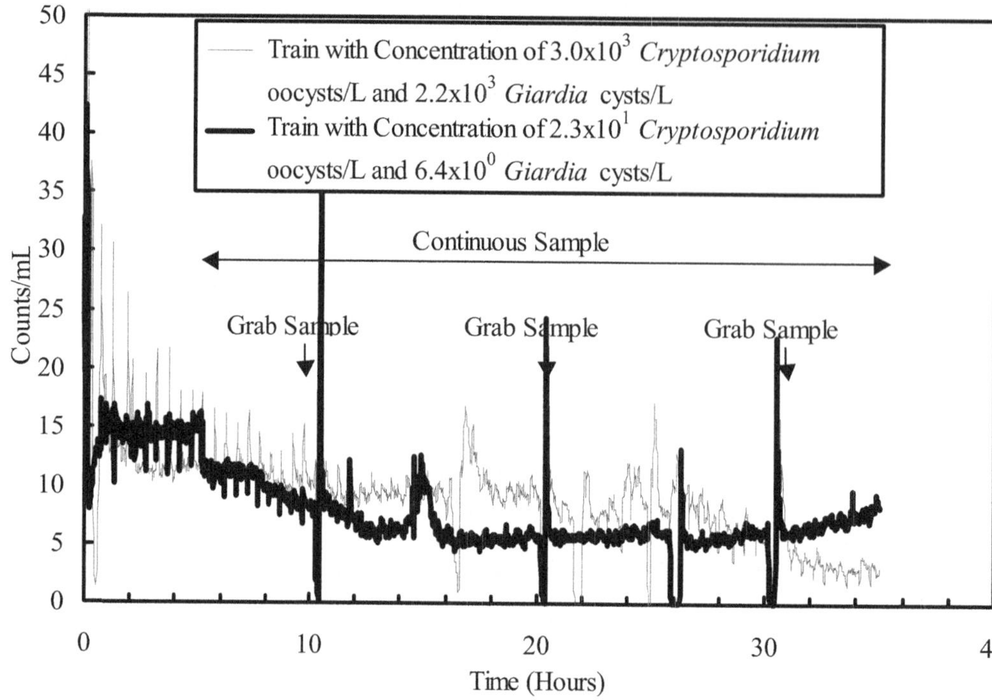

Figure D.4 10 to 15 μm size particles in filtration effluent versus run time in the March 2006 experiment.

the treatment train that received the lower pathogen dose. Similar results were observed for particles in the 10 to 15 μm size range, as shown in Figure D.4. This size range represents typical sizes of *Giardia* cysts.

Pathogen Removal through Conventional Treatment for Filter Effluent Grab Samples

Table D.2 shows the averaged results for filter effluent grab samples based on both microscopy and FC sorting-microscopy results. With initial concentrations of 3.0 x 10^3 oocysts/L and 2.2 x 10^3 cysts/L, log removals of 1.6 ± 0.1 and 2.7 ± 0.1 were obtained for oocysts and cysts, respectively. There were no *Cryptosporidium* oocysts or *Giardia* cysts found in samples taken after filtration for the train with spike doses of 2.3 x 10^1 oocysts/L and 6.4 x 10^0 cysts/L. Log removals for these samples were calculated from the mean log number of pathogens in raw water samples and log of detection limit.

Pathogen Removal through Conventional Treatment for Continuous Samples Collected from Filter Effluent

Log removal results from continuous samples were taken continuously over 30 hours after the ripening phase was complete. The sample times were noted on Figures D.2 through D.4. At higher initial spiking doses (3.0 x 10^3 oocysts/L and 2.2 x 10^3 cysts/L), log removals of 1.8 and 2.8 were observed for *Cryptosporidium* oocysts and *Giardia* cysts, respectively. No cysts or oocysts were detected in the filter effluent when the lower spike train (2.3 x 10^1

Table D.2
Pathogen removal through conventional treatment for the March 2006 experiment.

Filter Run Time (hours)	*Cryptosporidium* Log Removal		*Giardia* Log Removal	
	Spike Dose = 3.0×10^3 oocysts/L	Spike Dose = 2.3×10^1 oocysts/L	Spike Dose = 2.2×10^3 cysts/L	Spike Dose = 6.4×10^0 cysts/L
10	1.6	> 1.8	2.6	> 1.3
20	1.6	> 1.8	2.8	> 1.3
30	1.5	> 1.8	2.6	> 1.3

oocysts/L and 6.4×10^0 cysts/L) was analyzed by FC sorting-microscopy. Thus, log removals for these samples were calculated from the mean log number of pathogens in raw water samples and log of detection limit. The log removals more than 2.3 and 1.8 were observed.

Pathogen Removal from Sedimentation through Granular Media Filtration

Pathogen Removal through Filtration from Filter Effluent Grab Samples

When the higher pathogen concentrations at the sedimentation effluent were 1.5×10^2 oocysts/L and 6.9×10^1 cysts/L, filtration achieved average log removals of 0.3 ± 0.1 and 1.2 ± 0.1, respectively. In the lower concentration treatment train, pathogen concentrations in the settled water were below detection limit. Thus, log removal across filtration could not be calculated.

Pathogen Removal through Filtration from Filter Effluent Continuous Samples

Log removal results from continuous samples were obtained from spiked Lake Mendota water that went from sedimentation to filtration as part of the conventional treatment train. The samples were taken continuously for 30 hours as shown in Figures D.2 through D.4. For *Cryptosporidium* oocysts, log removals of 0.5 were achieved across filtration with a filter influent concentration of 1.5×10^2 oocysts/L. Filtration was able to remove 1.3 log of *Giardia* cysts when filter influent cyst concentrations were 6.9×10^1 cysts/L.

Comparison of Grab Sampling with Continuous Sampling

Figures D.5 and D.6 show a comparison of log removals based on grab sampling with those based on continuous sampling. For both *Cryptosporidium* oocysts and *Giardia* cysts, there was no significant difference between grab sampling and continuous sampling for this experiment.

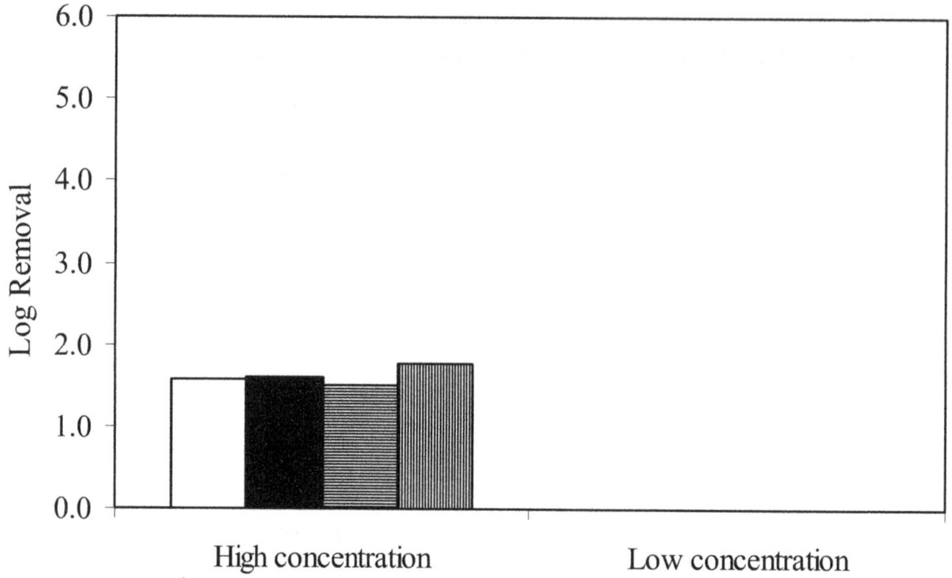

Figure D.5 Comparison of sampling methods for *Cryptosporidium* oocyst removal through conventional treatment. Symbols: □ Grab Sampling at 10 hours, ■ Grab Sampling at 20 hours, ⊟ Grab Sampling at 30 hours, ◫ Continuous Sampling from 5 to 35 hours.

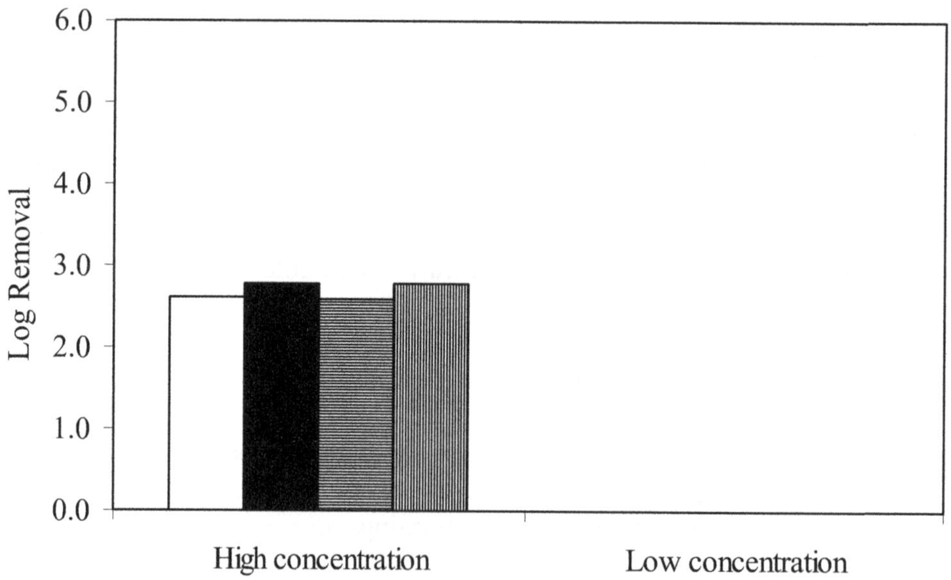

Figure D.6 Comparison of sampling methods for *Giardia* cyst removal through conventional treatment. Symbols: □ Grab Sampling at 10 hours, ■ Grab Sampling at 20 hours, ⊟ Grab Sampling at 30 hours, ◫ Continuous Sampling from 5 to 35 hours.

Cryptosporidium Oocyst Removal versus *Giardia* Cyst Removal

The results obtained from the March 2006 experiment were used to generate Figures D.7 to D.11. Figure D.7 compares log removal of *Cryptosporidium* oocysts with log removal of *Giardia* cysts through sedimentation. A paired *t*-test demonstrated that *Giardia* cysts were more readily removed than *Cryptosporidium* oocysts (p-value = 0.03). The mean log *Giardia* cyst removal was 0.2 log higher than the mean log *Cryptosporidium* oocyst removal. Figures D.8 and D.9 compare log removal of *Cryptosporidium* oocysts with log removal of *Giardia* cysts through conventional treatment. A paired *t*-test showed that *Giardia* cysts were more readily removed than *Cryptosporidium* oocysts at a 95% confidence level (p-value < 0.01). The mean log *Giardia* cyst removal was 1.1 log higher than the mean log *Cryptosporidium* oocyst removal. Figures D.10 and D.11 show the same comparison through filtration. A paired *t*-test demonstrated that the mean log *Giardia* cyst removal was significantly higher than the mean log *Cryptosporidium* oocyst removal (p-value < 0.01). The mean log *Giardia* cyst removal was 0.9 log higher than the mean log *Cryptosporidium* oocyst removal.

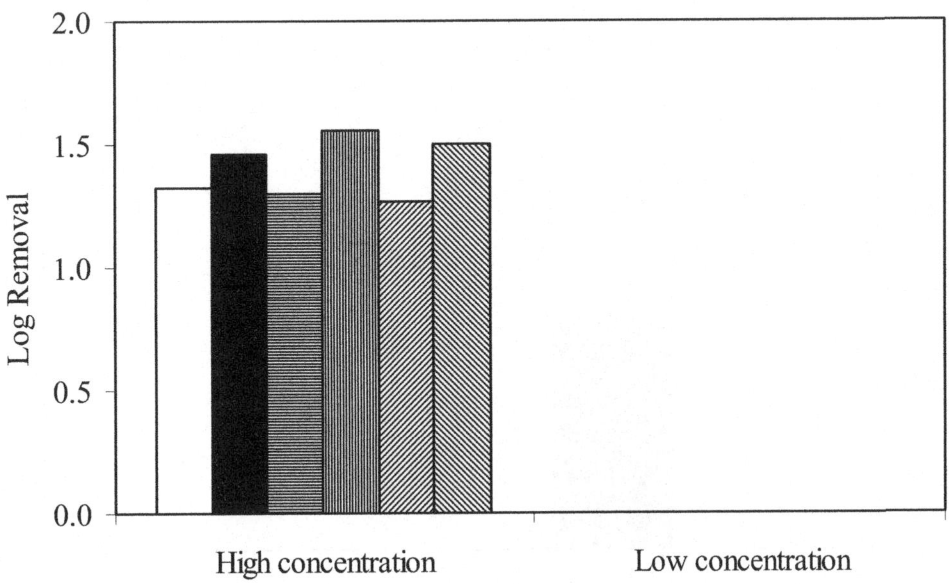

Figure D.7 Comparison of *Cryptosporidium* oocyst and *Giardia* cyst removals through sedimentation. Symbols: □ *Cryptosporidium* at 10 hours, ■ *Giardia* at 10 hours, ▤ *Cryptosporidium* at 20 hours, ▥ *Giardia* at 20 hours, ▨ *Cryptosporidium* at 30 hours, ▧ *Giardia* at 30 hours.

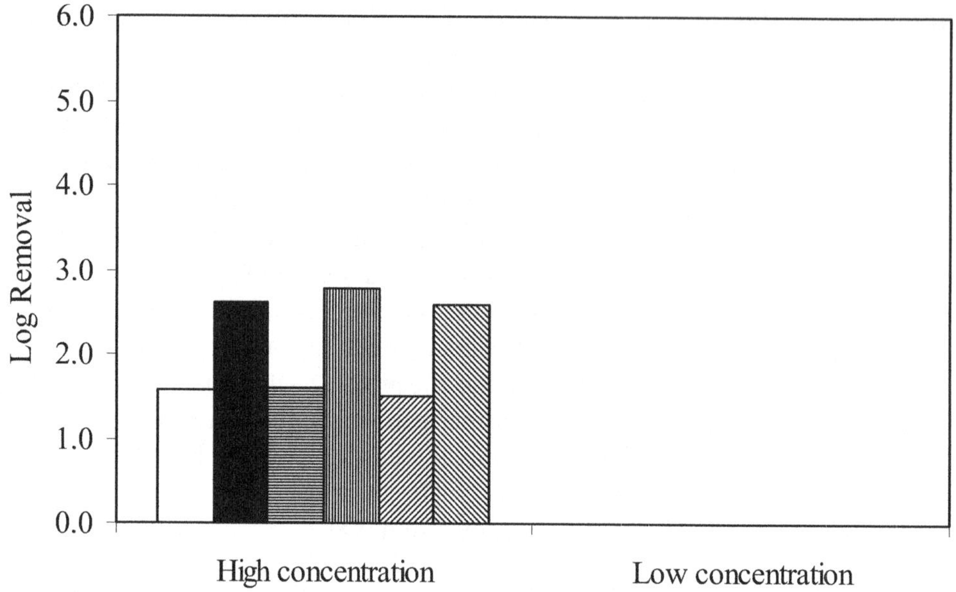

Figure D.8 **Comparison of *Cryptosporidium* oocyst and *Giardia* cyst removals through conventional treatment, based on grab samples. Symbols: □ *Cryptosporidium* at 10 hours, ■ *Giardia* at 10 hours, ▤ *Cryptosporidium* at 20 hours, ▥ *Giardia* at 20 hours, ▨ *Cryptosporidium* at 30 hours, ▧ *Giardia* at 30 hours.**

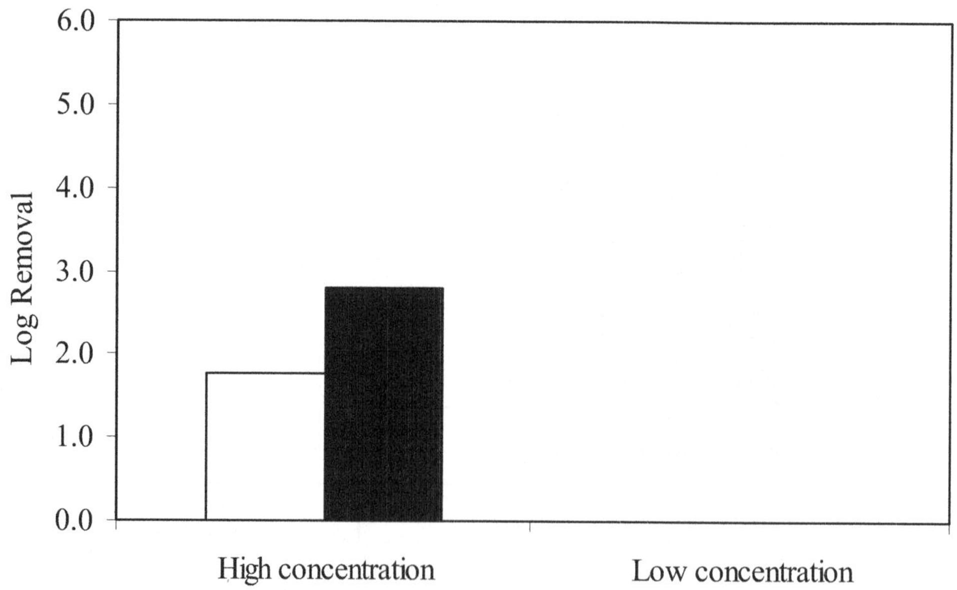

Figure D.9 **Comparison of *Cryptosporidium* oocyst and *Giardia* cyst removals through conventional treatment, based on continuous samples. Symbols: □ *Cryptosporidium*, ■ *Giardia*.**

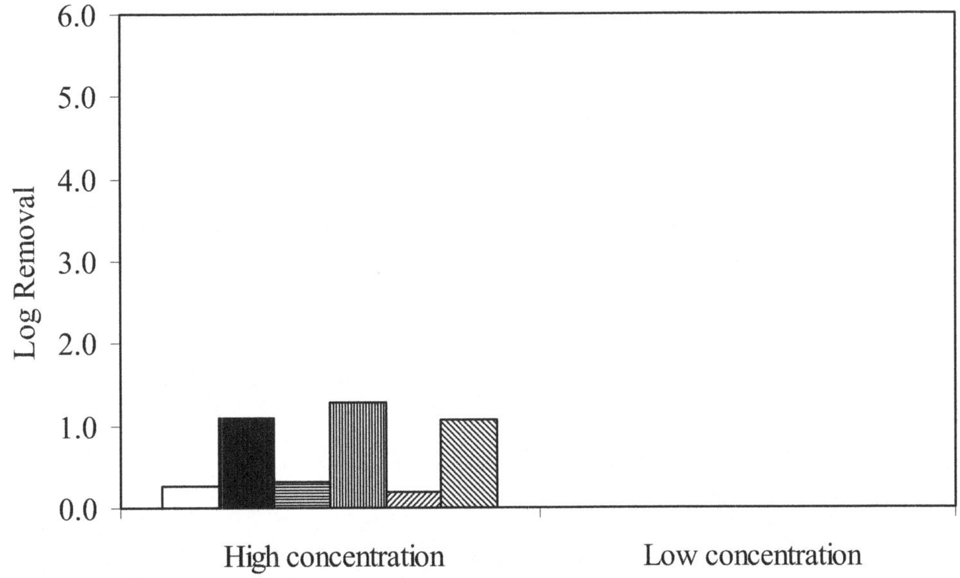

Figure D.10 Comparison of *Cryptosporidium* oocyst and *Giardia* cyst removals through filtration, based on grab samples. Symbols: □ *Cryptosporidium* at 10 hours, ■ *Giardia* at 10 hours, ▤ *Cryptosporidium* at 20 hours, ▥ *Giardia* at 20 hours, ▨ *Cryptosporidium* at 30 hours, ▧ *Giardia* at 30 hours.

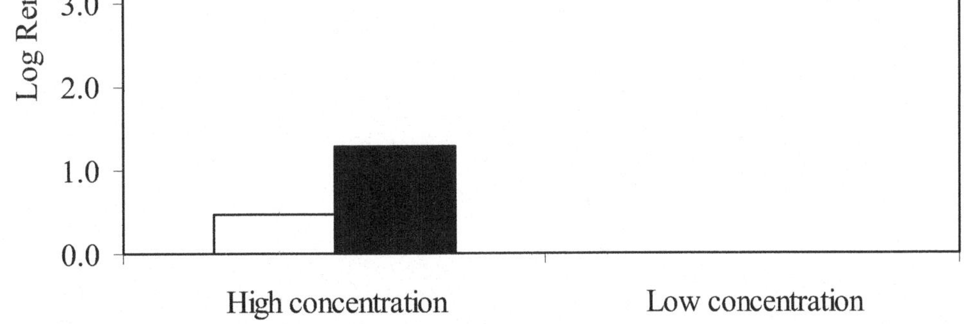

Figure D.11 Comparison of *Cryptosporidium* oocyst and *Giardia* cyst removals through filtration, based on continuous samples. Symbols: □ *Cryptosporidium*, ■ *Giardia*.

APPENDIX E
EXPERIMENTAL CONDITIONS AND RESULTS FOR THE DECEMBER 2006 EXPERIMENT

The alum dose for the treatment train (Train 1) with initial concentration of $2.7 \times 10^5 \pm 1.2 \times 10^5$ *Cryptosporidium* oocysts/L and $3.2 \times 10^5 \pm 1.3 \times 10^5$ *Giardia* cysts/L was set to 80 mg/L. The other treatment train (Train 2), spiked with $2.6 \times 10^1 \pm 1.3 \times 10^1$ *Cryptosporidium* oocysts/L and $2.4 \times 10^1 \pm 1.6 \times 10^1$ *Giardia* cysts/L, had an alum dose of 70 mg/L. The alum dose was based on jar tests where the source water was supplemented with test pathogens. This was done the night before the experiment to ensure that the selected pilot-scale coagulant dose accounted for the concentration of the pathogen feed. Therefore, both treatment trains were expected to run with optimal coagulation conditions. Raw water temperature during the experiment was 12.0 ± 1.1°C. Spiked raw water turbidity averaged 3.5 ± 1.0 ntu for the high concentration train and 3.6 ± 1.2 ntu for the low concentration train during the experiment. These raw water turbidities were significantly higher than observed in other experiments and were likely due to strong northwesterly winds over the lake. These winds kicked up after the alum dose was established and placed into operation.

Pathogen Removal from Alum Addition through Sedimentation

Turbidity Results

Turbidity in the sedimentation effluent averaged 0.58 ± 0.09 ntu for the high concentration train and 0.81 ± 0.20 ntu for the low concentration train (see Figure E.1). These results show that the low concentration train had periodic stability problems, possibly caused by suboptimal coagulation conditions resulting from the strong winds. In both treatment trains, treatment reduced UV254 by 55 percent from $0.08\ cm^{-1}$ to $0.04\ cm^{-1}$.

Pathogen Removal through Sedimentation from Grab Samples

Table E.1 shows the effect of spike dose on pathogen removal through sedimentation. Log removals of 1.2 ± 0.1 and 1.3 ± 0.1 were achieved for oocysts and cysts, respectively, with initial concentrations of 2.7×10^5 oocysts/L and 3.2×10^5 cysts/L. With initial concentrations of 2.6×10^1 oocysts/L and 2.4×10^1 cysts/L, there were no *Cryptosporidium* oocysts or *Giardia* cysts detected in the majority of settled water samples. As mentioned in Appendix C, log removals for these samples were calculated from the mean log number of pathogens in raw water samples and log average of detection limit.

Pathogen Removal from Alum Addition through Granular Media Filtration

Turbidity Results

Filter effluent turbidity data are shown in Figure E.2. The average filter effluent turbidity was 0.09 ± 0.01 ntu for the train with the higher pathogen dose and was 0.16 ± 0.02 ntu for the train with the lower pathogen dose. A paired *t*-test showed that the mean filter effluent turbidity

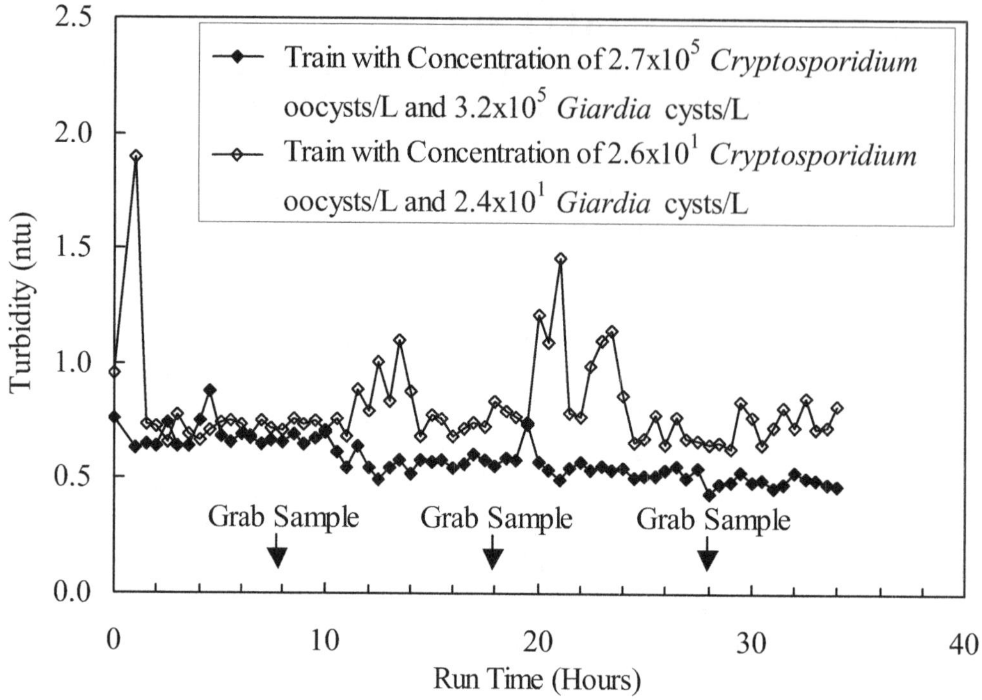

Figure E.1 Turbidity in sedimentation effluent versus run time.

Table E.1
Pathogen removal through sedimentation for the December 2006 experiment.

Filter Run Time (hours)	*Cryptosporidium* Log Removal		*Giardia* Log Removal	
	Spike Dose = 2.7 x 10^5 oocysts/L	Spike Dose = 2.6 x 10^1 oocysts/L	Spike Dose = 3.2 x 10^5 cysts/L	Spike Dose = 2.4 x 10^1 cysts/L
10	1.1	> 1.3	1.3	> 1.3
20	1.3	> 1.3	1.4	> 1.3
30	1.2	> 1.3	1.3	> 1.3

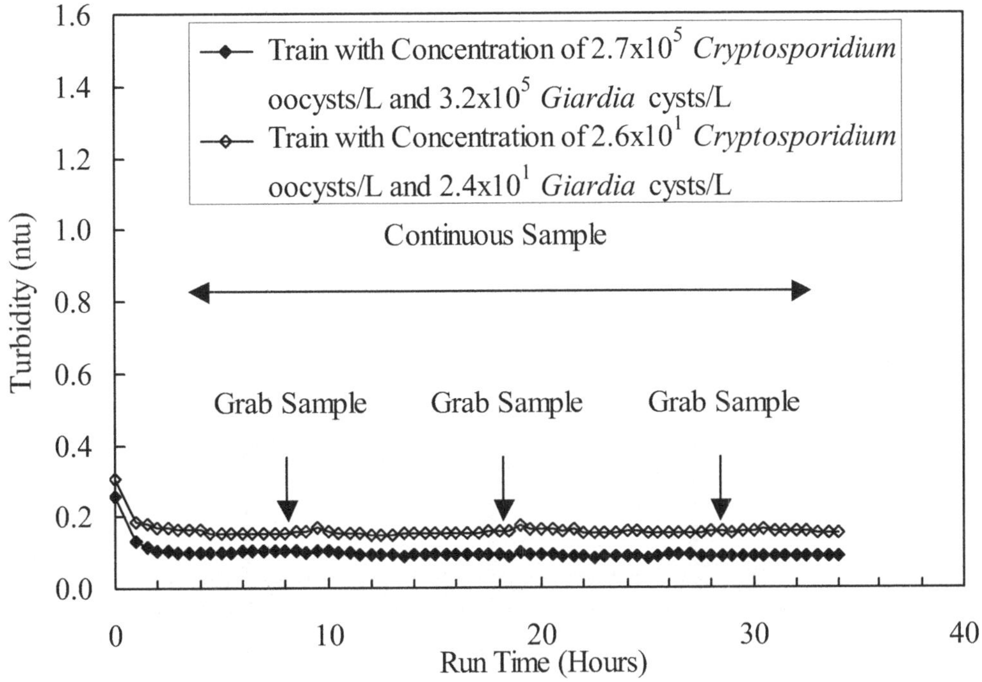

Figure E.2 Turbidity in filtration effluent versus time.

of the train with the higher pathogen dose was significantly different from the mean filter effluent turbidity of the train with lower pathogen dose (p-value < 0.01).

Particle Count Results

Filter effluent particle count data are shown in Figures E.3 and E.4 for 3 to 7 µm sized particles and for 10 to 15 µm sized particles, respectively. For the train with the higher pathogen dose, the particle count dropped to near zero after 20 hours of filter run time. This was due to a sample feed failure to this particle counter during the experiment. Therefore, the comparison between the two treatment trains was based on the average particle counts during the first 20 hours of filter run time. As shown in Figure E.3, the treatment train that received the higher pathogen dose had an average particle count of 3.7×10^1 particles/mL, which was comparable to the average particle count of 3.1×10^1 particles/mL in the treatment train that received the lower pathogen dose. For particles in the 10 to 15 µm size range, the train with the higher pathogen dose had an average particle count of 1.4×10^0 particles/mL, which was one order of magnitude higher than the train with the lower pathogen dose which had an average particle count of 2.9×10^{-1} particles/mL.

Pathogen Removal through Conventional Treatment for Filter Effluent Grab Samples

Table E.2 shows the averaged results for filter effluent grab samples based on both microscopy and FC sorting-microscopy results. With initial concentrations of 2.7×10^5 oocysts/L and 3.2×10^5 cysts/L, log removals of 4.2 ± 0.1 and 4.4 were obtained for oocysts and

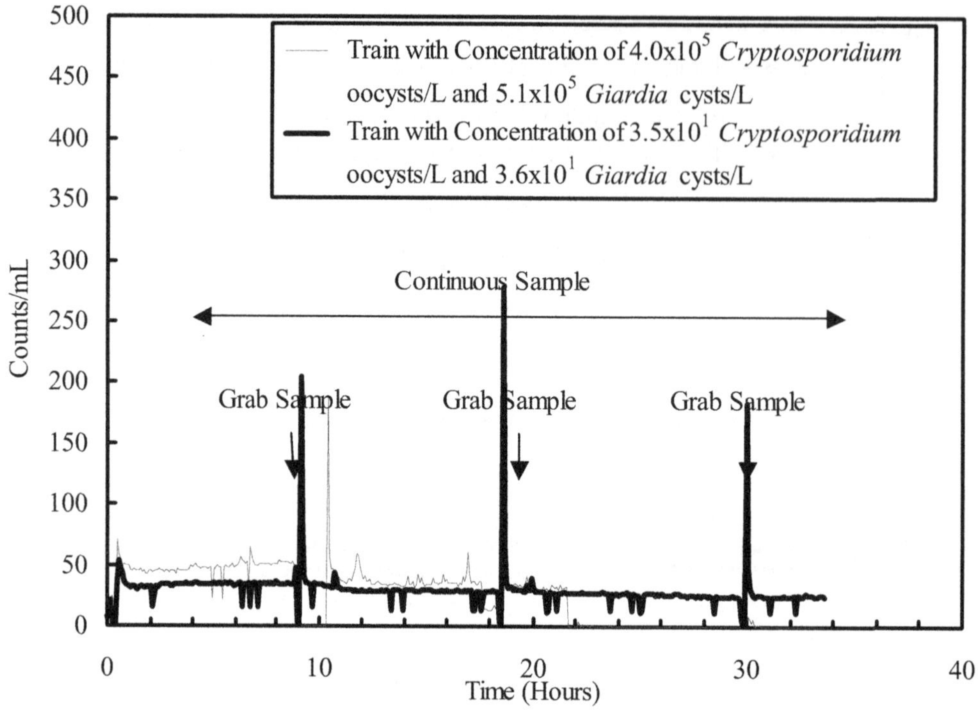

Figure E.3 3 to 7 µm size particles in filtration effluent versus run time in the December 2006 experiment.

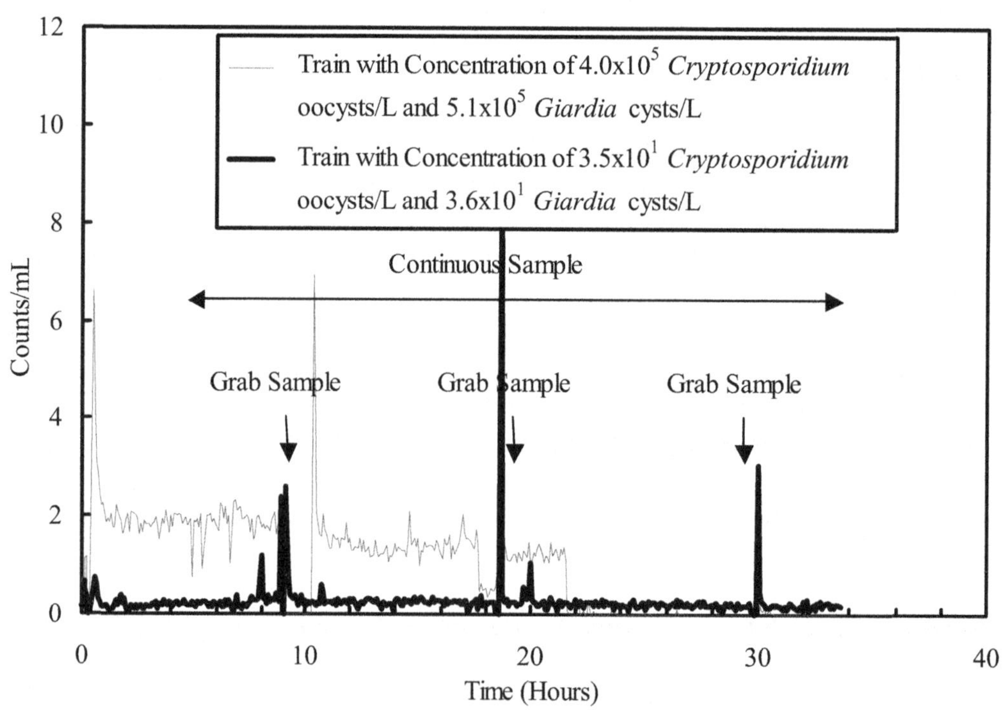

Figure E.4 10 to 15 µm size particles in filtration effluent versus run time in the December 2006 experiment.

Table E.2
Pathogen removal through conventional treatment for the December 2006 experiment.

Filter Run Time (hours)	*Cryptosporidium* Log Removal		*Giardia* Log Removal	
	Spike Dose = 2.7×10^5 oocysts/L	Spike Dose = 2.6×10^1 oocysts/L	Spike Dose = 3.2×10^5 cysts/L	Spike Dose = 2.4×10^1 cysts/L
10	4.1	> 1.8	4.4	> 1.8
20	4.3	> 1.1	>4.6	> 0.8
30	>4.2	> 1.8	>4.4	> 1.8

cysts, respectively. When filter effluent samples were analyzed by microscopy, there were no *Cryptosporidium* oocysts or *Giardia* cysts found in the train with spike doses of 2.6×10^1 oocysts/L and 2.4×10^1 cysts/L. Minimum log removals for these samples were calculated from the mean log number of pathogens in raw water samples and mean log of detection limit. These were averaged with log removals observed by FC sorting-microscopy and reported in Table E.2 with a ">" sign.

Pathogen Removal through Conventional Treatment for Continuous Samples Collected from Filter Effluent

Log removal results from continuous samples were taken continuously over 30 hours after the ripening phase was complete. The sample times were noted on Figures E.2 through E.4. At higher initial spiking doses (2.7×10^5 oocysts/L and 3.2×10^5 cysts/L), log removals of 5.2 and 5.4 were observed for *Cryptosporidium* oocysts and *Giardia* cysts, respectively. A log removal of 2.6 was observed at the lower oocyst spike dose (2.6×10^1 oocysts/L). There were no *Giardia* cysts found in the train with lower spike dose (2.4×10^1 cysts/L) analyzed by FC sorting-microscopy.

Pathogen Removal from Sedimentation through Granular Media Filtration

Pathogen Removal through filtration from Filter Effluent Grab Samples

Log removal of 3.0 ± 0.1 oocysts/L and 3.0 cysts/L were observed for the train with higher pathogen concentrations in the settled water. In the lower concentration train, pathogen concentrations after sedimentation were below the detection limit, so log removal could not be calculated.

Pathogen Removal through filtration for Continuous Samples Collected from Filter Effluent

Log removal through filtration was observed from continuous sampling. Again, the samples were taken continuously for 30 hours as shown in Figures E.2 through E.4. For *Cryptosporidium* oocysts, log removals of 4.0 were achieved across filtration when the filter influent concentration was 1.7 x 10^4 oocysts/L. Filtration was able to remove 4.0 log of *Giardia* cysts when filter influent cyst concentrations were 1.5 x 10^4 cysts/L. As with grab sampling, log removal could not be calculated in the lower pathogen concentration train because settled water pathogen counts were below detection limit.

Comparison of Grab Sampling with Continuous Sampling

Figures E.5 and E.6 show a comparison of log removals based on grab sampling with those based on continuous sampling. The observed *Cryptosporidium* oocyst removals based on grab sampling at high initial concentration of pathogens were about one log lower than the observed removal based on continuous sampling. At low initial oocyst concentration, continuous sampling achieved detectable concentrations while grab sampling did not. For *Giardia* cyst, both continuous and grab sampling results were under detection limit.

Cryptosporidium Oocyst Removal versus *Giardia* Cyst Removal

The results obtained from the December 2006 experiment were used to generate Figures E.7 through E.11. Figure E.7 compares log removal of *Cryptosporidium* oocysts with log removal of *Giardia* cysts through sedimentation. A paired *t*-test demonstrated that the log removal of *Giardia* cysts were not significantly different from log *Cryptosporidium* oocyst removal at a 95 % confidence level (p-value = 0.12). Figures E.8 and E.9 compare log removal of *Cryptosporidium* oocysts with log removal of *Giardia* cysts through conventional treatment. The log *Giardia* cyst removal was slightly higher than the log *Cryptosporidium* oocyst removal.

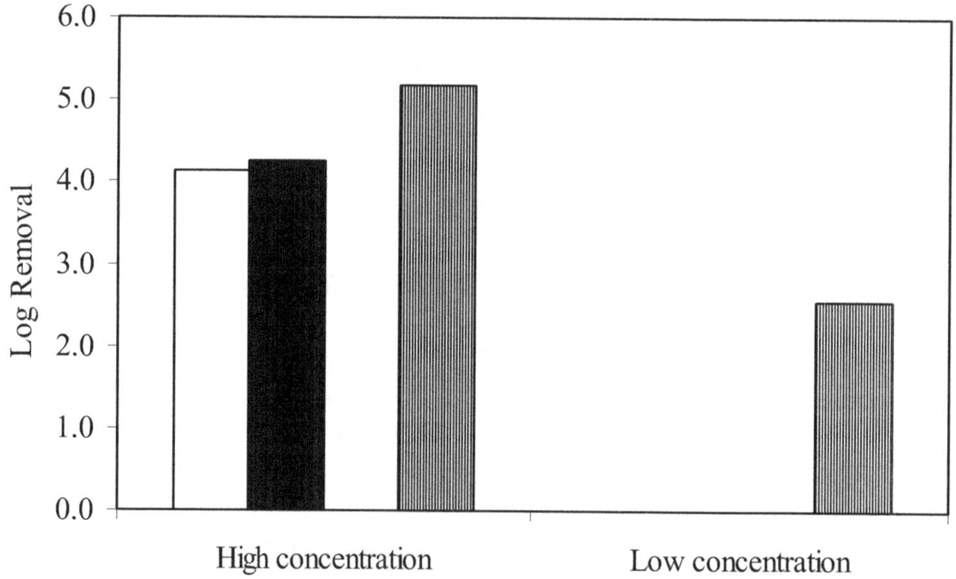

Figure E.5 Comparison of sampling methods for *Cryptosporidium* oocyst removal through conventional treatment. Symbols: □ Grab Sampling at 10 hours, ■ Grab Sampling at 20 hours, ⊟ Grab Sampling at 30 hours, ▥ Continuous Sampling from 5 to 35 hours.

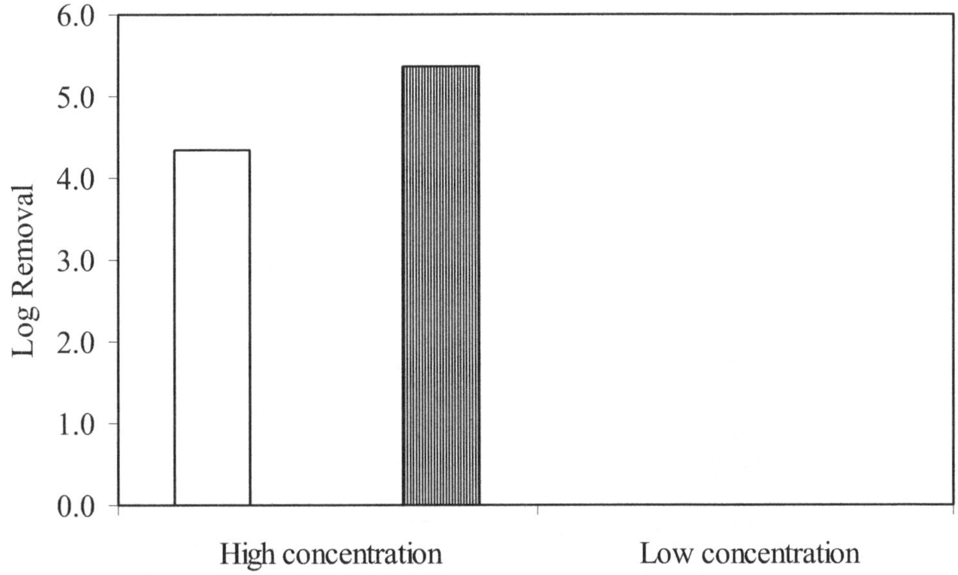

Figure E.6 **Comparison of sampling methods for *Giardia* cyst removal through conventional treatment. Symbols: □ Grab Sampling at 10 hours, ■ Grab Sampling at 20 hours, ▤ Grab Sampling at 30 hours, ▥ Continuous Sampling from 5 to 35 hours.**

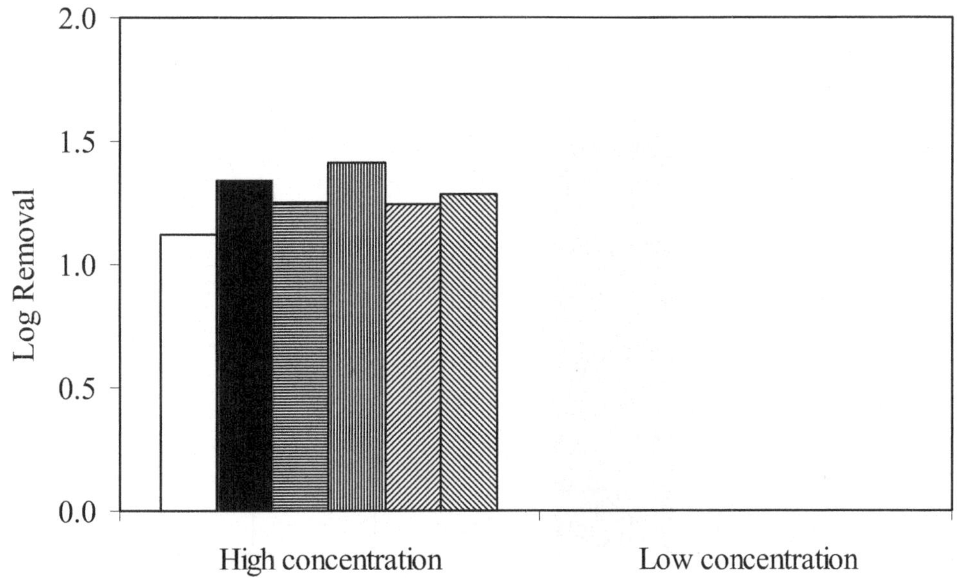

Figure E.7 **Comparison of *Cryptosporidium* oocyst and *Giardia* cyst removals through sedimentation. Symbols: □ *Cryptosporidium* at 10 hours, ■ *Giardia* at 10 hours, ▤ *Cryptosporidium* at 20 hours, ▥ *Giardia* at 20 hours, ▨ *Cryptosporidium* at 30 hours, ▧ *Giardia* at 30 hours.**

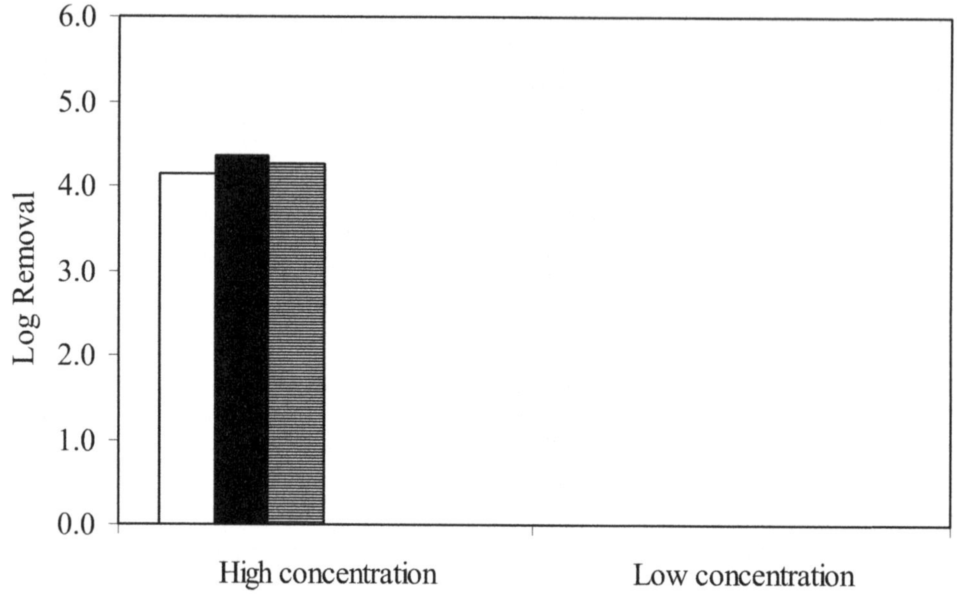

Figure E.8 Comparison of *Cryptosporidium* oocyst and *Giardia* cyst removals through conventional treatment, based on grab samples. Symbols: □ *Cryptosporidium* at 10 hours, ■ *Giardia* at 10 hours, ▤ *Cryptosporidium* at 20 hours, ▥ *Giardia* at 20 hours, ▨ *Cryptosporidium* at 30 hours, ▧ *Giardia* at 30 hours.

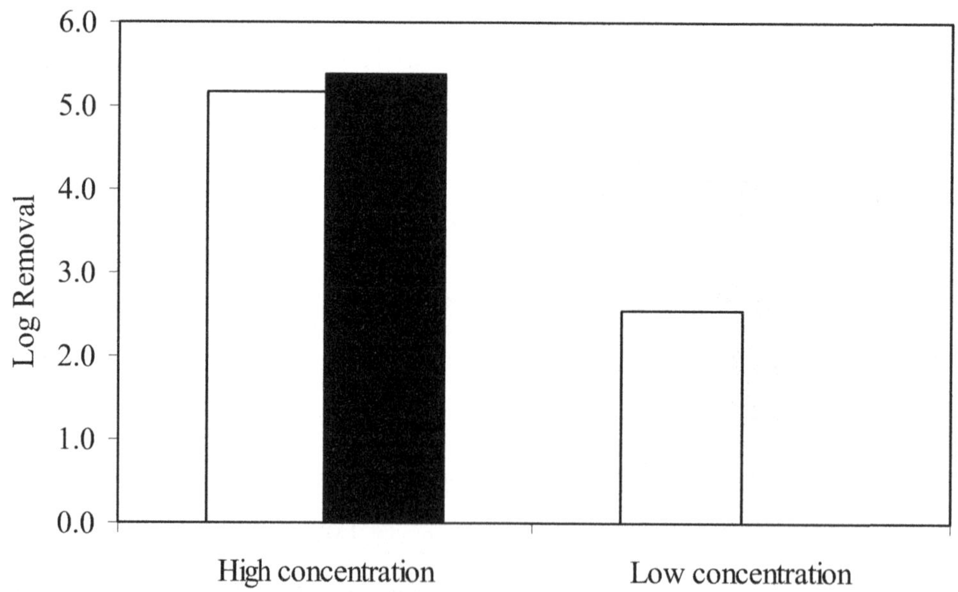

Figure E.9 Comparison of *Cryptosporidium* oocyst and *Giardia* cyst removals through conventional treatment, based on continuous samples. Symbols: □ *Cryptosporidium*, ■ *Giardia*.

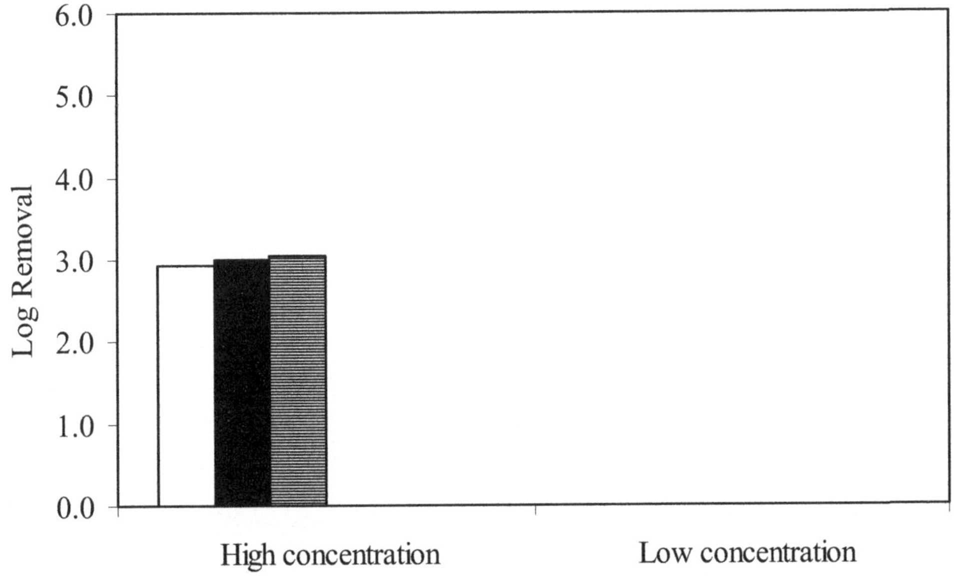

Figure E.10 Comparison of *Cryptosporidium* oocyst and *Giardia* cyst removals through filtration, based on grab samples. Symbols: □ *Cryptosporidium* at 10 hours, ■ *Giardia* at 10 hours, ▤ *Cryptosporidium* at 20 hours, ▥ *Giardia* at 20 hours, ▨ *Cryptosporidium* at 30 hours, ▧ *Giardia* at 30 hours.

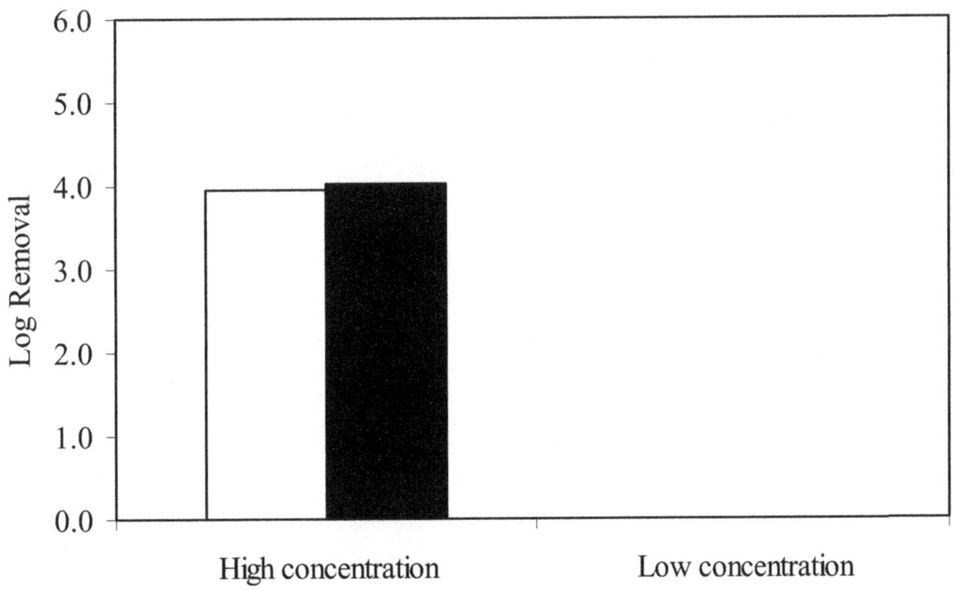

Figure E.11 Comparison of *Cryptosporidium* oocyst and *Giardia* cyst removals through filtration, based on continuous samples. Symbols: □ *Cryptosporidium*, ■ *Giardia*.

Figures E.10 and E.11 show the same comparison through filtration. The log *Giardia* cyst removals show slightly higher than the log *Cryptosporidium* oocyst removals.

APPENDIX F
EXPERIMENTAL CONDITIONS AND RESULTS FOR THE MARCH 2007 EXPERIMENT

The alum dose for the treatment train (Train 1) with initial concentration of $1.7 \times 10^1 \pm 1.2 \times 10^1$ *Cryptosporidium* oocysts/L and $1.8 \times 10^1 \pm 1.5 \times 10^1$ *Giardia* cysts/L was set to 70 mg/L. The other treatment train (Train 2), spiked with $2.3 \times 10^5 \pm 1.1 \times 10^5$ *Cryptosporidium* oocysts/L and $2.4 \times 10^5 \pm 1.4 \times 10^5$ *Giardia* cysts/L, had an alum dose of 80 mg/L. The alum dose was based on jar tests where the source water was supplemented with test pathogens. This was done to ensure that the selected pilot-scale coagulant dose accounted for the concentration of the pathogen feed. Therefore, both treatment trains were run with optimal coagulation conditions. Raw water temperature during the experiment was 8.0 ± 0.0°C. Spiked raw water turbidity averaged 1.19 ± 0.17 ntu for the low concentration train and 1.22 ± 0.13 ntu for the high concentration train during the experiment.

Pathogen Removal from Alum Addition through Sedimentation

Turbidity Results

Turbidity in the sedimentation effluent averaged 0.79 ± 0.18 ntu for the low spike train and 0.63 ± 0.13 ntu for the high spike train (see Figure F.1). These results show that both treatment trains had periodic stability problems. In treatment train 1, treatment reduced UV254

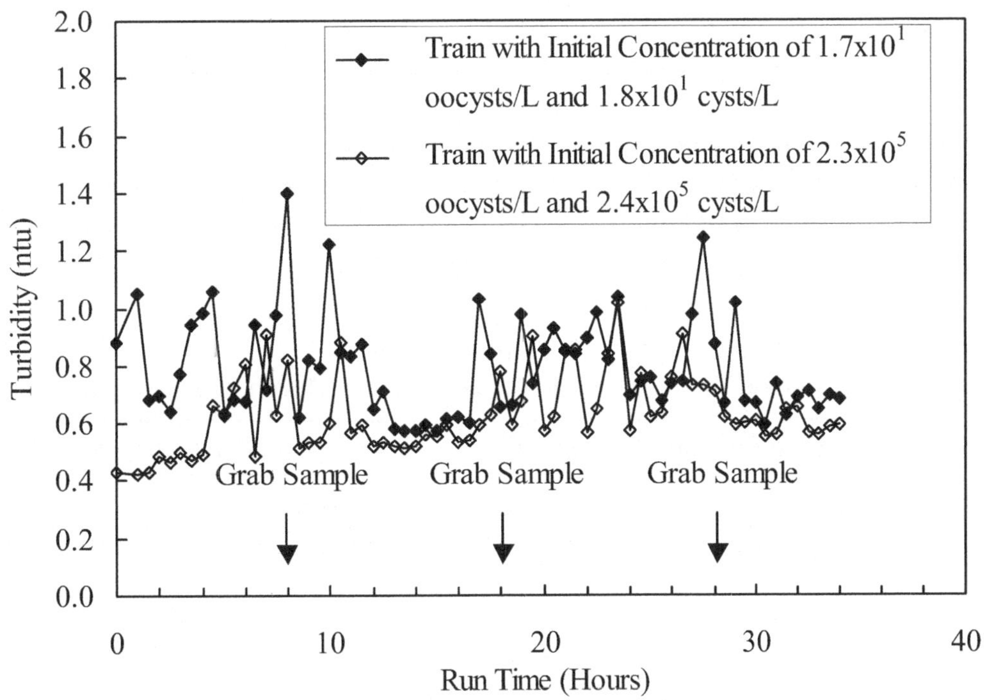

Figure F.1 Turbidity in sedimentation effluent versus run time.

by 40 percent from 0.07 cm^{-1} to 0.04 cm^{-1} while treatment train 2 reduced UV254 by 44 percent from 0.05 cm^{-1} to 0.03 cm^{-1}

Pathogen Removal through Sedimentation from Grab Samples

Table F.1 shows the effect of spike dose on pathogen removal through sedimentation. Log removals of 1.3 ± 0.1 and 1.4 ± 0.1 were achieved for oocysts and cysts, respectively, with initial concentrations of 2.3 x 10^5 oocysts/L and 2.4 x 10^5 cysts/L. With initial concentrations of 1.7 x 10^1 oocysts/L and 1.8 x 10^1 cysts/L, there were no *Cryptosporidium* oocysts or *Giardia* cysts found in the settled water. As mentioned in Appendix C, log removals for these samples were calculated from the mean log number of pathogens in raw water samples and the mean log of the detection limit.

Pathogen Removal from Alum Addition through Granular Media Filtration

Turbidity Results

Filter effluent turbidity data are shown in Figure F.2. The average filter effluent turbidity was 0.06 ± 0.01 ntu for the train with the higher pathogen dose and was 0.10 ± 0.01 ntu for the train with the lower pathogen dose. A paired *t*-test showed that the mean filter effluent turbidity of the train with the higher pathogen dose was significantly different from the mean filter effluent turbidity of the train with lower pathogen dose (p-value < 0.01).

Particle Count Results

Filter effluent particle count data are shown in Figures F.3 and F.4 for 3 to 7 µm sized particles and for 10 to 15 µm sized particles, respectively. As shown in Figure F.3, the treatment train that received the higher pathogen dose had an average particle count of 3.6 x 10^1 particles/mL, which was one order of magnitude lower than the average particle count of 1.8 x 10^2 particles/mL in the treatment train that received the lower pathogen dose. Similar results were observed for particles in the 10 to 15 µm size range, as shown in Figure F.4. This size range represents typical sizes of *Giardia* cysts. The results agreed with turbidity results that the

Table F.1
Pathogen removal through sedimentation for the March 2007 experiment.

Filter Run Time (hours)	*Cryptosporidium* Log Removal		*Giardia* Log Removal	
	Spike Dose = 2.3 x 10^5 oocysts/L	Spike Dose = 1.7 x 10^1 oocysts/L	Spike Dose = 2.4 x 10^5 cysts/L	Spike Dose = 1.8 x 10^1 cysts/L
10	1.3	> 0.6	1.4	> 0.7
20	1.3	> 0.6	1.4	> 0.7
30	1.4	> 0.6	1.5	> 0.7

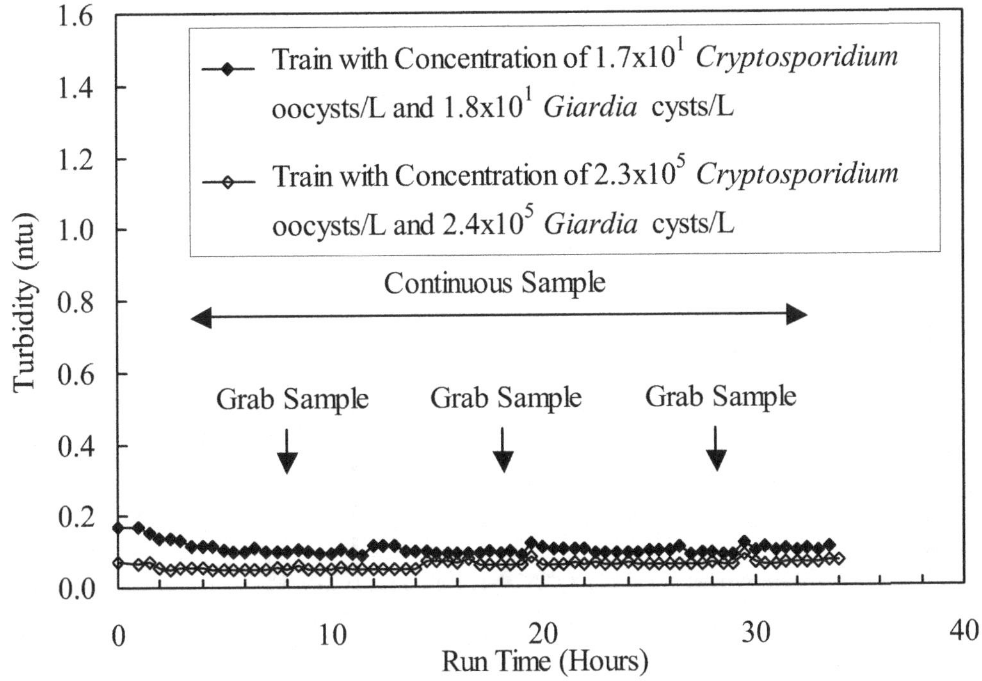

Figure F.2 Turbidity in filtration effluent versus time.

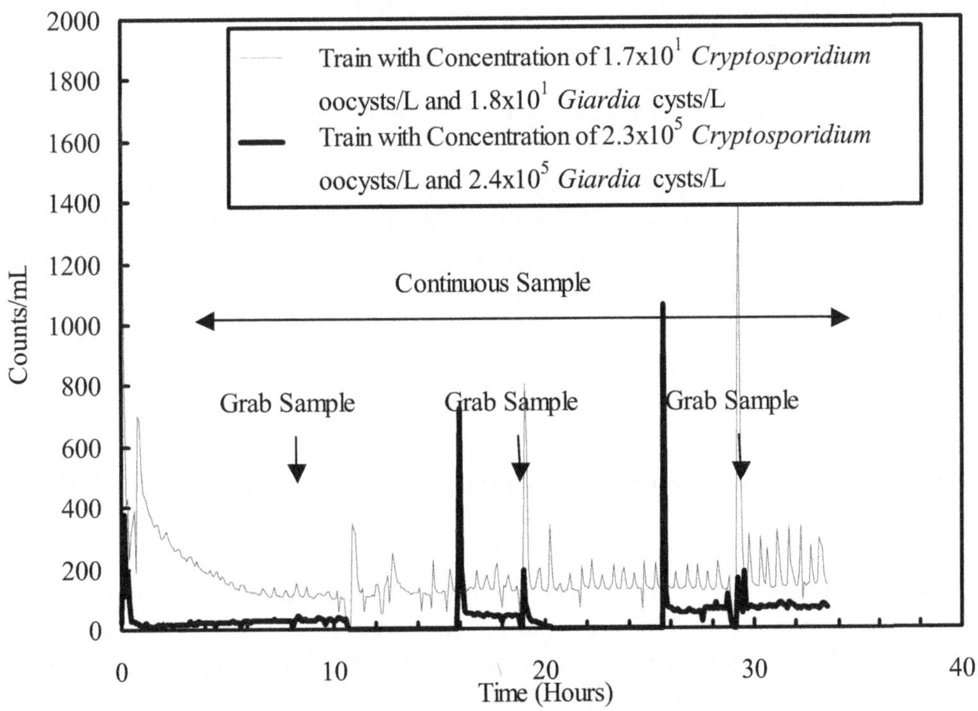

Figure F.3 3 to 7 μm size particles in filtration effluent versus run time in the March 2007 experiment.

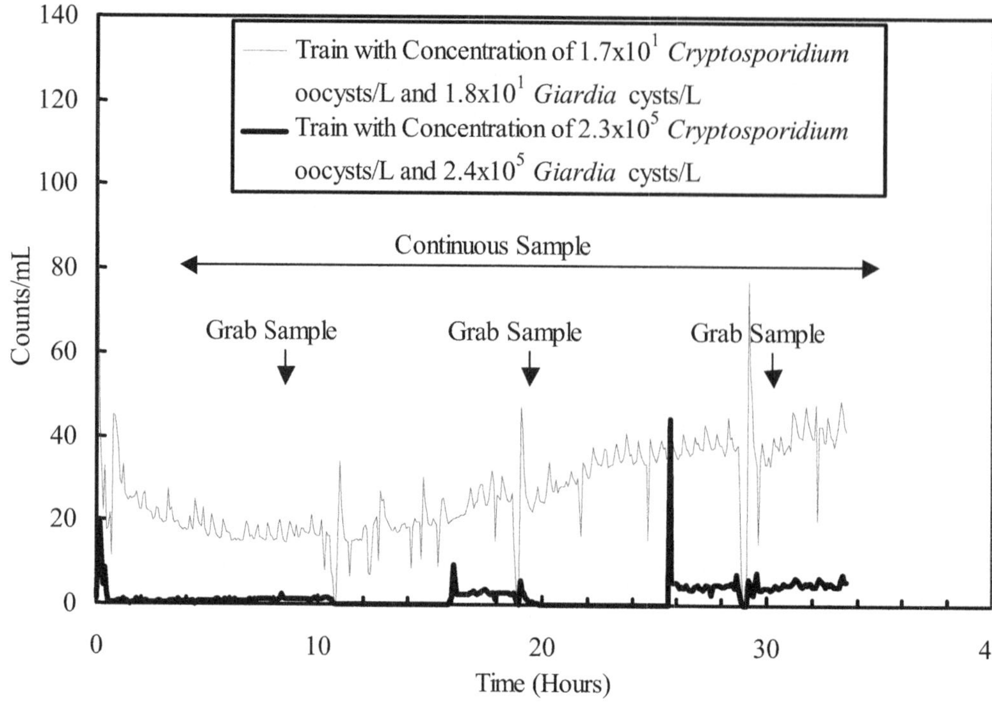

Figure F.4 10 to 15 µm size particles in filtration effluent versus run time in the March 2007 experiment.

treatment train with lower pathogen concentration had higher turbidity than the treatment train with higher pathogen concentration.

Pathogen Removal through Conventional Treatment for Filter Effluent Grab Samples

Table F.2 shows the averaged results for filter effluent grab samples based on both microscopy and FC sorting-microscopy results. With initial concentrations of 2.3 x 10^5 oocysts/L and 2.4 x 10^5 cysts/L, log removals of 2.6 ± 0.2 and 3.0 ± 0.1 were obtained for oocysts and cysts, respectively. There were no *Cryptosporidium* oocysts or *Giardia* cysts found in samples taken after filtration for the train with spike doses of 1.7 x 10^1 oocysts/L and 1.8 x 10^1 cysts/L. Log removals for these samples were calculated from the mean log number of pathogens in raw water samples and log of detection limit.

Pathogen Removal through Conventional Treatment for Continuous Samples Collected from Filter Effluent

Log removal results from continuous samples were taken continuously over 30 hours after the ripening phase was complete. The sample times were noted on Figures F.2 through F.4. At higher initial spiking doses (2.3 x 10^5 oocysts/L and 2.4 x 10^5 cysts/L), log removals of 2.7 and 3.4 were observed for *Cryptosporidium* oocysts and *Giardia* cysts, respectively. Log removals of 2.2 and 2.5 were detected in the filter effluent at the treatment train with the lower spike doses (2.3 x 10^1 oocysts/L and 6.4 x 10^0 cysts/L).

Table F.2
Pathogen removal through conventional treatment for the March 2007 experiment.

Filter Run Time (hours)	*Cryptosporidium* Log Removal		*Giardia* Log Removal	
	Spike Dose = 2.3×10^5 oocysts/L	Spike Dose = 1.7×10^1 oocysts/L	Spike Dose = 2.4×10^5 cysts/L	Spike Dose = 1.8×10^1 cysts/L
10	2.8	> 1.6	3.1	> 1.7
20	2.6	> 1.6	3.1	> 1.7
30	2.4	> 1.5	2.9	> 1.7

Pathogen Removal from Sedimentation through Granular Media Filtration

Pathogen Removal through filtration from Filter Effluent Grab Samples

For the treatment train with the higher pathogen concentrations in the settled water were 1.1×10^4 oocysts/L and 9.1×10^3 cysts/L, filtration achieved average log removals of 1.3 ± 0.2 and 1.6 ± 0.1, respectively. In the lower concentration train, pathogen concentrations after sedimentation were below detection limit and log removal across filtration could not be calculated.

Pathogen Removal through filtration for Continuous Samples Collected from Filter Effluent

Log removal results from continuous samples were obtained from spiked Lake Mendota water that went from sedimentation to filtration as part of conventional treatment. The samples were taken continuously for 30 hours as shown in Figures F.2 through F.4. For *Cryptosporidium* oocysts, log removals of 1.3 were achieved across filtration with initial concentrations of 1.1×10^4 oocysts/L. Filtration was able to remove 2.0 log of *Giardia* cysts when cyst concentrations were 9.1×10^3 cysts/L.

Comparison of Grab Sampling with Continuous Sampling

Figures F.5 and F.6 show a comparison of log removals based on grab sampling with those based on continuous sampling. The observed *Cryptosporidium* oocyst removals based on grab sampling at high initial concentration of pathogens were about the same as the observed removal based on continuous sampling. For *Giardia* cysts, the log removal based on continuous sampling were slightly higher than the log removals based on grab sampling in the high concentration train. At low initial concentration of pathogens, continuous sampling produced detectable concentrations while grab sampling did not.

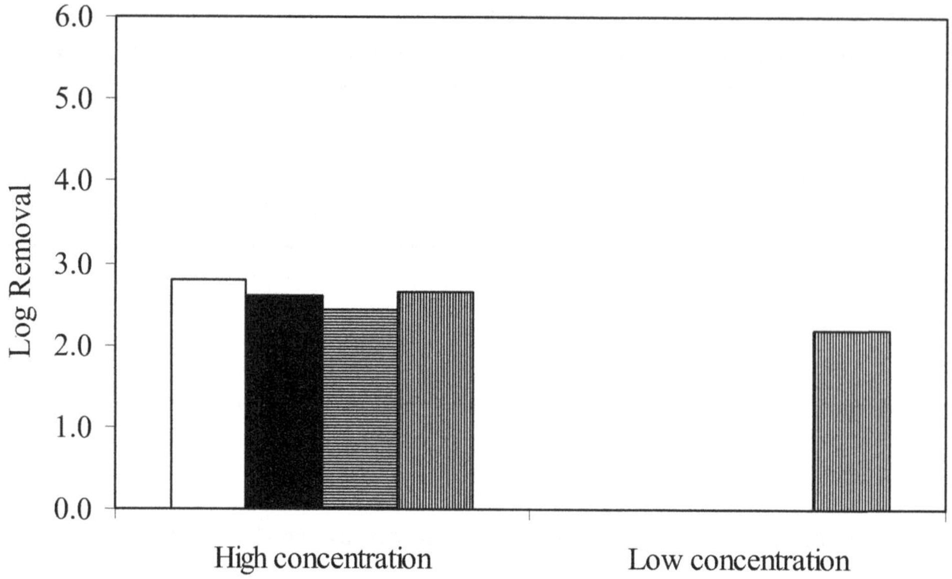

Figure F.5 Comparison of sampling methods for *Cryptosporidium* oocyst removal through conventional treatment. Symbols: □ Grab Sampling at 10 hours, ■ Grab Sampling at 20 hours, ▤ Grab Sampling at 30 hours, ▥ Continuous Sampling from 5 to 35 hours.

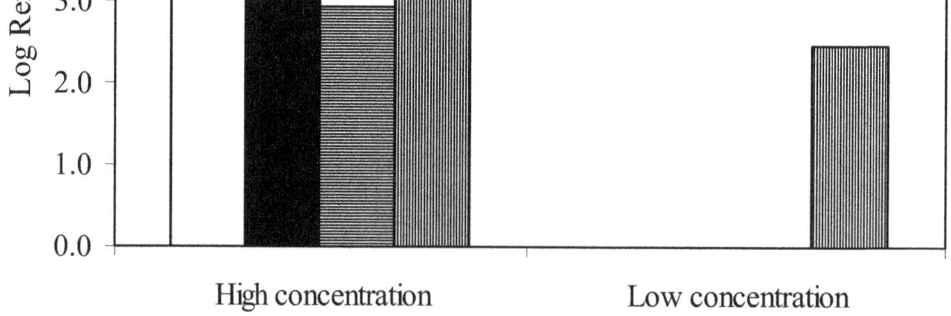

Figure F.6 Comparison of sampling methods for *Giardia* cyst removal through conventional treatment. Symbols: □ Grab Sampling at 10 hours, ■ Grab Sampling at 20 hours, ▤ Grab Sampling at 30 hours, ▥ Continuous Sampling from 5 to 35 hours.

Cryptosporidium Oocyst Removal versus *Giardia* Cyst Removal

The results obtained from the March 2007 experiment were used to generate Figures F.7 to F.11. Figure F.7 compares log removal of *Cryptosporidium* oocysts with log removal of *Giardia* cysts through sedimentation. A paired *t*-test demonstrated that *Giardia* cysts removals were not significantly different from *Cryptosporidium* oocysts removals (p-value = 0.16). Figures F.8 and F.9 compare log removal of *Cryptosporidium* oocysts with log removal of *Giardia* cysts through conventional treatment. A paired *t*-test showed that *Giardia* cysts were more readily removed than *Cryptosporidium* oocysts at a 95% confidence level (p-value = 0.02). The mean log *Giardia* cyst removal was 0.5 log higher than the mean log *Cryptosporidium* oocyst removal. Figures F.10 and F.11 show the same comparison through filtration. A paired *t*-test demonstrated that the mean log *Giardia* cyst removal was significantly higher than the mean log *Cryptosporidium* oocyst removal (p-value = 0.02). The mean log *Giardia* cyst removal was 0.4 log higher than the mean log *Cryptosporidium* oocyst removal.

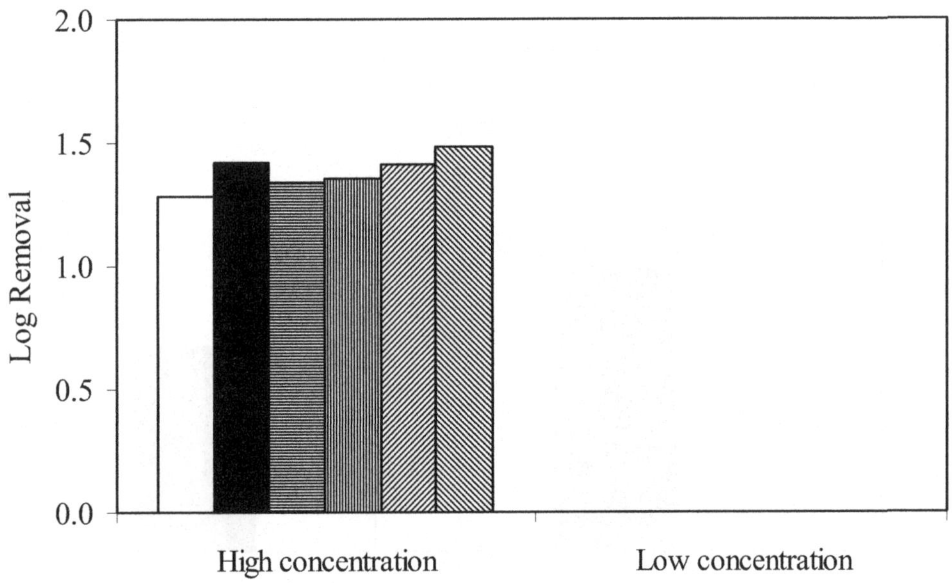

Figure F.7 Comparison of *Cryptosporidium* oocyst and *Giardia* cyst removals through sedimentation. Symbols: □ *Cryptosporidium* at 10 hours, ■ *Giardia* at 10 hours, ▤ *Cryptosporidium* at 20 hours, ▥ *Giardia* at 20 hours, ▨ *Cryptosporidium* at 30 hours, ▧ *Giardia* at 30 hours.

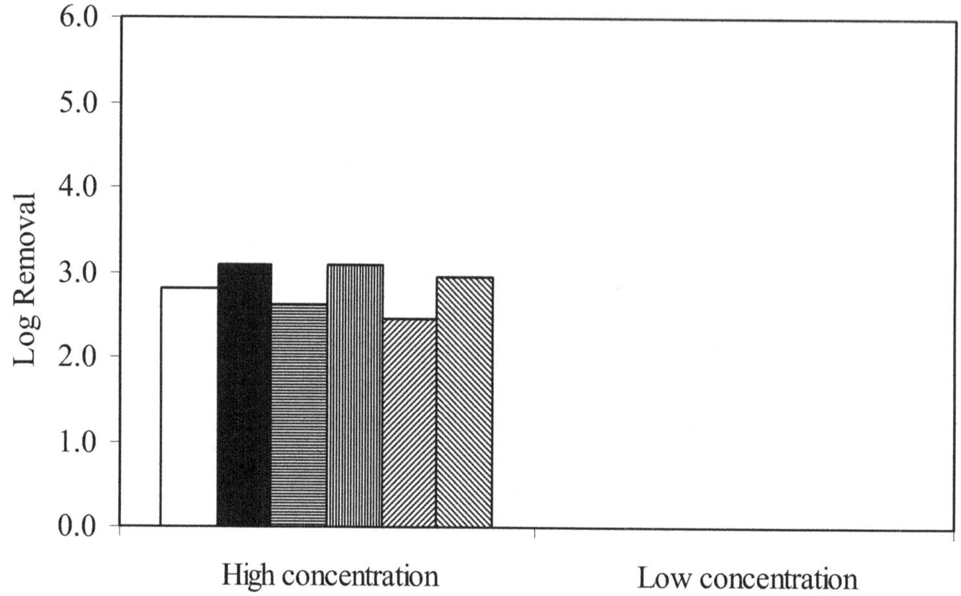

Figure F.8 **Comparison of *Cryptosporidium* oocyst and *Giardia* cyst removals through conventional treatment, based on grab samples. Symbols: □ *Cryptosporidium* at 10 hours, ■ *Giardia* at 10 hours, ▤ *Cryptosporidium* at 20 hours, ▥ *Giardia* at 20 hours, ▨ *Cryptosporidium* at 30 hours, ▧ *Giardia* at 30 hours.**

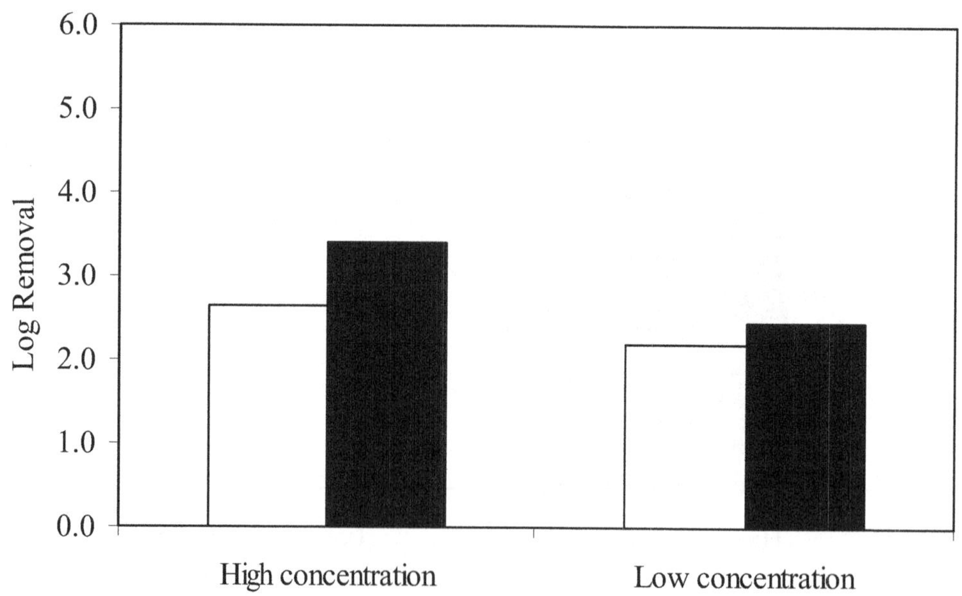

Figure F.9 **Comparison of *Cryptosporidium* oocyst and *Giardia* cyst removals through conventional treatment, based on continuous samples. Symbols: □ *Cryptosporidium*, ■ *Giardia*.**

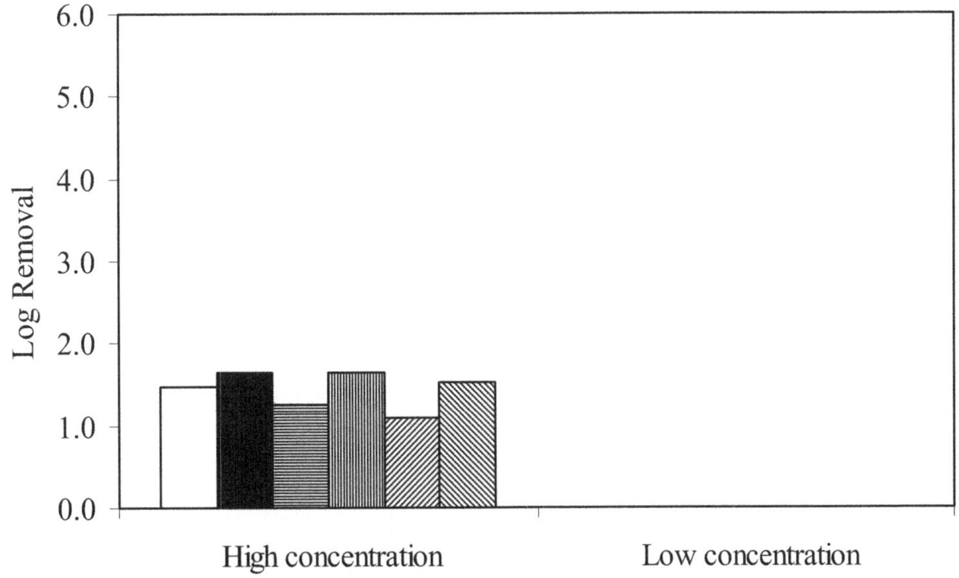

Figure F.10 Comparison of *Cryptosporidium* oocyst and *Giardia* cyst removals through filtration, based on grab samples. Symbols: □ *Cryptosporidium* at 10 hours, ■ *Giardia* at 10 hours, ▤ *Cryptosporidium* at 20 hours, ▥ *Giardia* at 20 hours, ▨ *Cryptosporidium* at 30 hours, ▧ *Giardia* at 30 hours.

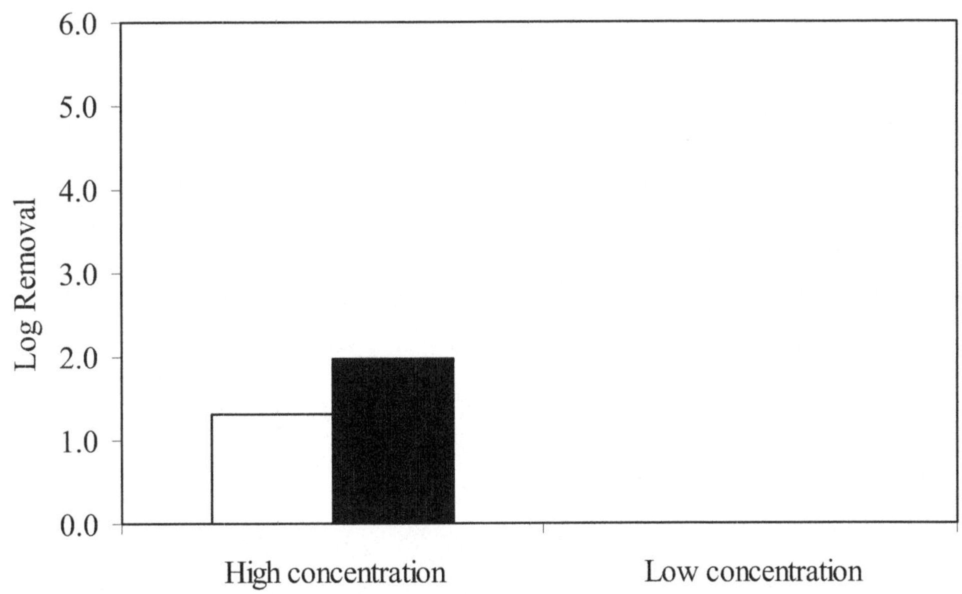

Figure F.11 Comparison of *Cryptosporidium* oocyst and *Giardia* cyst removals through filtration, based on continuous samples. Symbols: □ *Cryptosporidium*, ■ *Giardia*.

APPENDIX G
RECOVERY EFFICIENCY FOR THE PATHOGEN REMOVAL EXPERIMENTS

Recovery controls and matrix background checks were run at the onset of each experiment. For recovery controls, standards containing 500 oocysts and 500 cysts were seeded into 10 L grab samples of unseeded matrix water. These correspond to concentrations of 5×10^1 oocysts/L and 5×10^1 cysts/L. The standards were prepared by personnel at the Wisconsin State Laboratory of Hygiene by using FC sorting-microscopy. Three different matrix waters were tested: raw water, settled water and filtered water. These waters were collected at the onset of each experiment, before spiking pathogens into both treatment trains. After adding the standards to the matrix water, the samples were then processed and pathogens were detected using the procedures described in the main body of this report (see page 13 for a description of the two enumeration methods).

Table G.1 lists the pathogen counts from all six experiments. Because the concentrated sample was divided into four aliquots, each was expected to contain 125 oocysts and 125 cysts. The number of counted oocysts or cysts was then divided by the seed number of 125 to calculate the recovery efficiency. Figures G.1 and G.2 show the recovery efficiency results for samples analyzed by microscopy. The recovery efficiency results for samples analyzed by FC sorting-microscopy are shown in Figures G.3 and G.4.

With respect to background check, no *Cryptosporidium* oocysts or *Giardia* cysts were found in the 288 L of unseeded Lake Mendota water collected during each of the six experiments.

Table G.1
Pathogen counts in matrix spikes from all six experiments.

Date	Location	*Cryptosporidium* Oocyst Count by FC Sorting-Microscopy	*Cryptosporidium* Oocyst Count by Microscopy	*Giardia* Cyst Count by FC Sorting-Microscopy	*Giardia* Cyst Count by Microscopy
November 04	Raw	54	42	54	36
	Settled	na	56	na	56
	Filtered	107	63	81	66
January 05	Raw	36	94	47	84
	Settled	67	86	47	94
	Filtered	78	88	25	98
March 05	Raw	101	62	178	74
	Settled	na	60	na	30
	Filtered	8	92	27	104
March 06	Raw	117	90	86	79
	Settled	14	85	15	70
	Filtered	80	80	75	91
December 06	Raw	153	94	61	83
	Settled	7	85	7	95
	Filtered	41	86	66	92
March 07	Raw	0	83	566	86
	Settled	27	85	13	85
	Filtered	26	73	50	77

Note: the expected count was 125 pathogens.

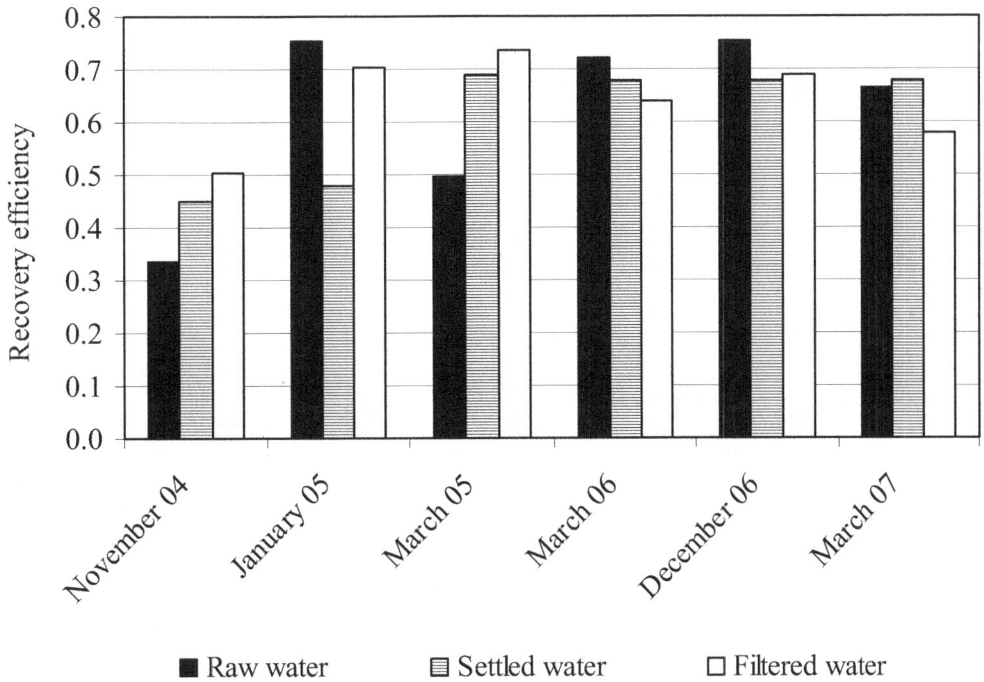

Figure G.1 ***Cryptosporidium*** **recovery efficiency analyzed by microscopy.**

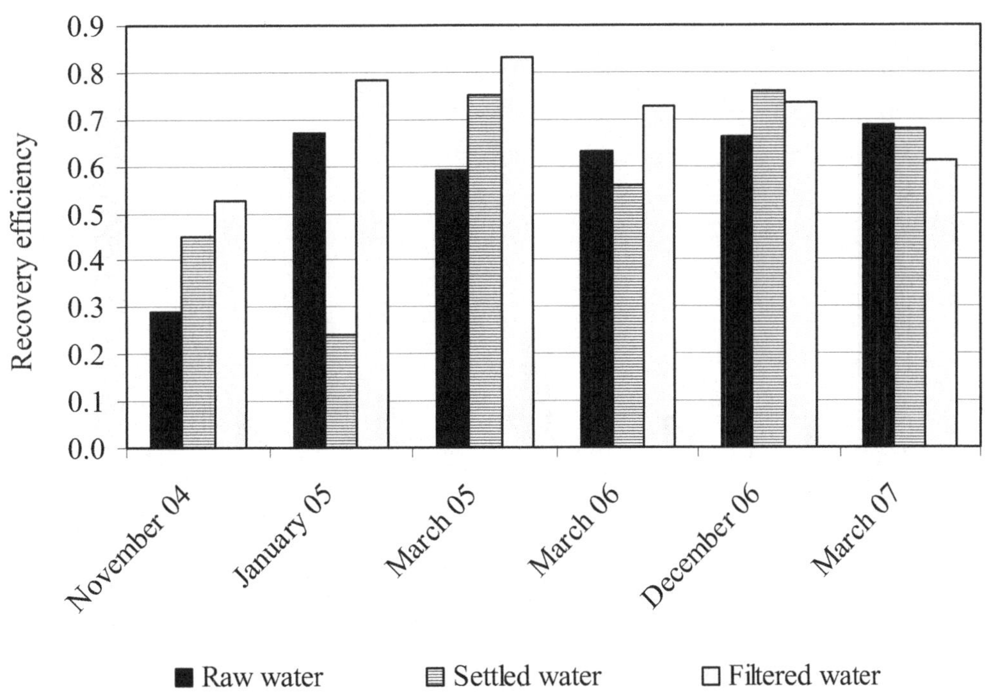

Figure G.2 ***Giardia*** **recovery efficiency analyzed by microscopy.**

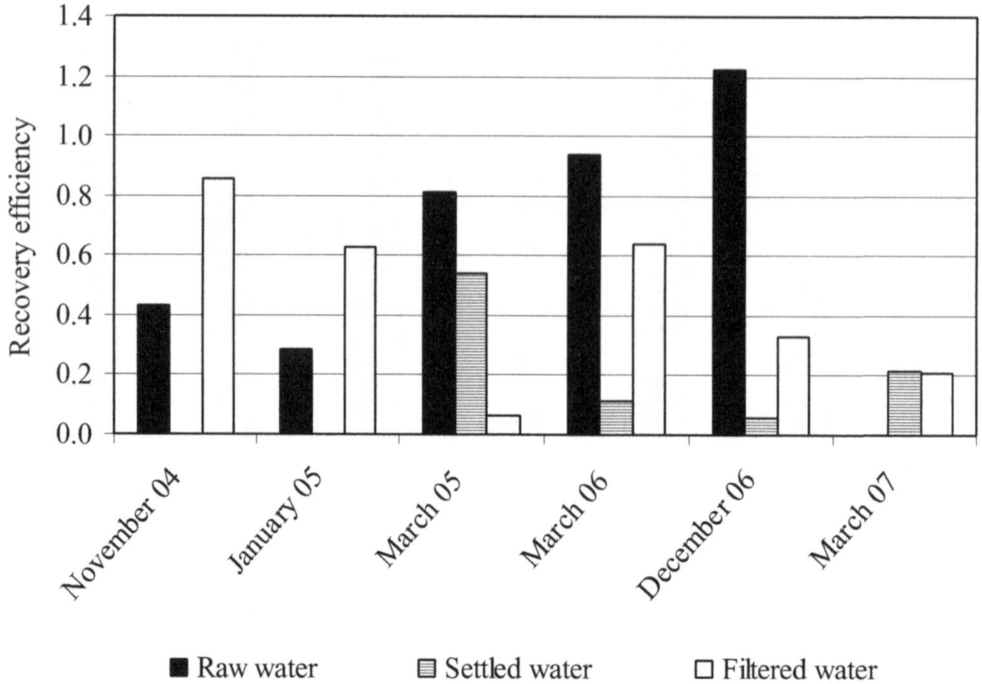

Figure G.3 ***Cryptosporidium*** **recovery efficiency analyzed by FC sorting-microscopy.**

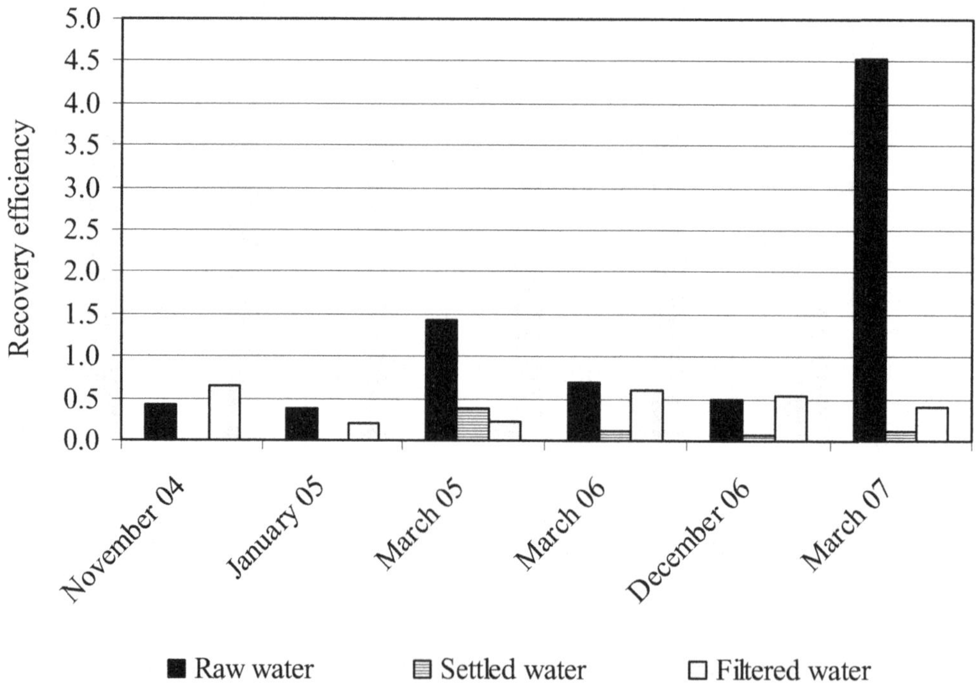

Figure G.4 ***Giardia*** **recovery efficiency analyzed by FC sorting-microscopy.**

APPENDIX H
CONCENTRATION FACTORS FOR THE PATHOGEN PELLETS ANALYZED BY FC-SORTING MICROSCOPY

Tables H.1 through H.6 show the concentration factors obtained by centrifugation of samples prior to analysis by flow cytometric sorting and microscopic enumeration.

Table H.1
Concentration factors for the pathogen pellets (November 2004 experiments).

Train	Location	Collection Method	Filter Run Time	Concentration Factors
1	raw	grab	10	94
		grab	20	150
		grab	30	100
	Settled	grab	10	na
		grab	20	na
		grab	30	na
	Filtered	grab	10	250
		grab	20	250
		grab	30	250
		continuous		250
2	Raw	grab	10	200
		grab	20	170
		grab	30	140
	Settled	grab	10	na
		grab	20	na
		grab	30	na
	Filtered	grab	10	250
		grab	10	250
		grab	20	250
		grab	20	250
		grab	30	250
		continuous		250

Table H.2
Concentration factors for the pathogen pellets (January 2005 experiments).

Train	Location	Collection Method	Filter Run Time	Concentration Factors
1	raw	grab	10	360
		grab	20	370
		grab	30	290
	Settled	grab	10	na
		grab	20	na
		grab	30	na
	Filtered	grab	10	250
		grab	10	250
		grab	20	250
		grab	30	250
		continuous		na
2	Raw	grab	10	400
		grab	20	330
		grab	30	310
	Settled	grab	10	na
		grab	20	na
		grab	30	na
	Filtered	grab	10	250
		grab	20	250
		grab	30	250
		continuous		360

Table H.3
Concentration factors for the pathogen pellets (March 2005 experiments).

Train	Location	Collection Method	Filter Run Time	Concentration Factors
1	raw	grab	10	260
		grab	20	240
		grab	30	300
	Settled	grab	10	na
		grab	10	na
		grab	20	na
		grab	30	na
	Filtered	grab	10	500
		grab	20	500
		grab	30	500
		continuous		500
2	Raw	grab	10	290
		grab	20	420
		grab	20	380
		grab	30	500
	Settled	grab	10	na
		grab	20	na
		grab	30	na
	Filtered	grab	10	500
		grab	10	500
		grab	20	500
		grab	30	500
		continuous		na

Table H.4
Concentration factors for the pathogen pellets (March 2006 experiments).

Train	Location	Collection Method	Filter Run Time	Concentration Factors
1	raw	grab	10	250
		grab	20	240
		grab	30	230
	Settled	grab	10	250
		grab	10	na
		grab	20	na
		grab	30	na
	Filtered	grab	10	250
		grab	20	250
		grab	30	250
		continuous		250
2	Raw	grab	10	230
		grab	10	210
		grab	20	270
		grab	30	230
	Settled	grab	10	190
		grab	20	170
		grab	30	na
	Filtered	grab	10	250
		grab	20	250
		grab	30	250
		continuous		260

Table H.5
Concentration factors for the pathogen pellets (December 2006 experiments).

Train	Location	Collection Method	Filter Run Time	Concentration Factors
1	raw	grab	10	61
		grab	10	120
		grab	20	110
		grab	30	68
	Settled	grab	10	120
		grab	20	110
		grab	30	96
	Filtered	grab	10	100
		grab	20	120
		grab	30	130
		continuous		75
2	Raw	grab	10	190
		grab	20	300
		grab	30	250
	Settled	grab	10	110
		grab	20	120
		grab	30	130
		grab	30	120
	Filtered	grab	10	250
		grab	20	250
		grab	30	200
		continuous		130

Table H.6
Concentration factors for the pathogen pellets (March 2007 experiments).

Train	Location	Collection Method	Filter Run Time	Concentration Factors
1	raw	grab	10	330
		grab	20	330
		grab	30	290
		grab	30	330
	Settled	grab	10	130
		grab	20	110
		grab	30	220
	Filtered	grab	10	500
		grab	20	500
		grab	30	630
		continuous		420
2	Raw	grab	10	240
		grab	20	220
		grab	30	280
	Settled	grab	10	250
		grab	10	200
		grab	20	77
		grab	30	83
	Filtered	grab	10	280
		grab	20	250
		grab	30	260
		continuous		240

REFERENCES

Al-Ani, M.Y., Hendricks, D.W., Logsdon, G.S., and Hibler, C.P. (1986) Removing *Giardia* Cysts from Low Turbidity Waters by Rapid Rate Filtration. *J. AWWA*. 78(5), 66-73.

Amirtharajah A. 1988. Some Theoretical and Conceptual Views of Filtration. *Jour. Amer. Water Works Assoc.* 80(12):36-46.

APHA, AWWA, and WEF. 1999. *Standard Methods for the Examination of Water and Wastewater.* 20th ed. Washington, D.C.: APHA.

Borchardt M.A. and S.K. Spencer. 2002. Concentration of *Cryptosporidium*, Microsporidia, and other Water-Borne Pathogens by Continuous Separation Channel Centrifugation. *J. Appl. Microbiol.* 92:649-656.

Burger R., S. Eyjje, K.H. Karlsen, and K.A. Lie. 2000. Numerical Methods for the Simulation of the Settling of Flocculated Suspensions. *Chemical Engineering Journal* 80(1-3):91-104.

Chang J.S. 1989. *Mathematical modeling of transient behavior of deep bed filtration.* D. Engng. dissertation, Asian Institution of Technology, Bangkok, Thailand.

Devore J.L. 2000. *Probability and Statistics for Engineering and the Sciences.* California, Duxbury.

DeWalle F.B., J. Engeset, and W. Lawrence. 1984. *Removal of Giardia Lamblia Cysts by Drinking Water Treatment Plants.* U.S. Environmental Protection Agency, Municipal Environmental Research Laboratory, Cincinnati, OH.

Draper N.R. and H. Smith. 1998. *Applied Regression Analysis.* New York: Wiley-Interscience.

Dugan N.R. K.R. Fox, J.H. Owens, and R.J. Miltner. 2001. Controlling *Cryptosporidium* Oocysts Using Conventional Treatment. *Jour. Amer. Water Works Assoc.* 93(12):64-76.

Edzwald J.K., J.E. Tobiason, L.M. Parento, M.B. Kelley, G.S. Kaminski, H.J. Dunn, and P.B. Galant. 2000. *Giardia* and *Cryptosporidium* Removals by Clarification and Filtration Under Challenge Conditions. *Jour. Amer. Water Works Assoc.* 92(12):70-84.

Ekie-Jezewska M.L., F. Feuillebois N., N. Lecoq, K. Masmoudi, R. Anthore, F. Bostel., and E. Wajnryb. 1999. Hydrodynamic Interactions between Two Spheres at Contact. *Physical Review E.* 59(3):3182-3191 Part B.

Emelko M.B., P.M. Huck, I.P. Douglas, and J. Van Den Oever 2000. *Cryptosporidium* and Microsphere Removal During Low Turbidity End-of-Run and Early Breakthrough Filtration. *Proc. of the 2000 AWWA Water Quality Technology Conference.* Denver: AWWA.

Emelko M.B. 2001. *Removal of Cryptosporidium parvum by Granular Media Filtration.* PhD dissertation. University of Waterloo. Waterloo, Ontario, Canada.

Emelko M.B., K.W. Guenbel, I.P. Douglas, and J.V. Den Oever. 2001. Assessing *Cryptosporidium* Removal by Filtration: Impact of Seeded Concentration. *Proc. of the 2001 AWWA Water Quality Technology Conference.* Denver: AWWA.

Emelko M.B., P.M. Huck, and I.P. Douglas. 2003. *Cryptosporidium* and Microsphere Removal During Late in Cycle Filtration. *Jour. Amer. Water Works Assoc.* 95(5):173-181.

Ginn T.M., A. Amirtharajah, and P.R. Karr. 1992. Effects of Particle Detachment in Granular-Media Filtration. *Jour. Amer. Water Works Assoc.* 84(2):66-76.

Han M.Y. and D.F. Lawler. 1992. The (Relative) Insignificance of *G* in Flocculation. *Jour. Amer. Water Works. Assoc.* 84(10):79-91.

Harrington G.W., H.W. Chen, A.J. Harris, I. Xagoraraki, D.A. Battigelli, and J.H. Standridge. 2001. *Removal of Emerging Waterborne Pathogens.* Denver, CO: AwwaRF.

Hendricks D.W., W.F. Clunie, G.D. Sturbaum, D.A. Klein, T.L. Champlin, P. Kugrens, J. Hirsch, B. McCourt, G.R. Nordby, M.D. Sobsey, D.J. Hunt, and M.J. Allen. 2005. Filtration Removals of Microorganisms and Particles. *Jour. Environ. Engr.* 131(12):1621-1632.

Hoffman R.M. and C. Chauret. 2000. *Evaluation of Antibodies to Cryptosporidium and Giardia Using Flow Cytometry.* Denver, CO: AwwaRF.

Huck P.M., B.M. Coffey, and C.R. O'Melia. 2001. *Filter Operation Effects on Pathogen Passage.* Denver, CO: AwwaRF.

Hunt J.R. 1982. Self-Similar Particle Size Distributions During Coagulation: Theory and Experiment Verification. *Jour. Fluid Mech.* 122:169

Iwasaki T. 1937. Some Notes on Sand Filtration. *Jour. Amer. Water Works Assoc.* 29(10):1591-1602.

Lang J.S., J.J. Giron, A.T. Hansen, R.R. Trussell, and W.E. Hodges. 1993. Investigating Filter Performance as a Function of the Ratio of Filter Size to Media Size. *Jour. Amer. Water Works Assoc.* 85(10):122-130.

Lawler D.F., E. Izurieta, and C.J. Kao. 1983. Changes in Particle Size Distributions in Batch Flocculation. *Jour. Amer. Water Works Assoc.* 75:12:604.

LeChevallier M.W., W.D. Norton, and R.G. Lee. 1991. Occurrence of *Giardia* and *Cryptosporidium* spp. in Surface Water Supplies. *Appl. Environ. Microbiol.* 57:2610-2616.

LeChevallier M.W. and W.D. Norton. 1995. *Giardia* and *Cryptosporidium* in Raw and Finished Water. *Jour. Amer. Water Works Assoc.* 87(9):54-68.

LeChevallier, M.W., G.D. Di. Giovanni, J.L. Clancy, Z. Bukhari, S. Bukhari, J.S. Rosen, and M.M. Frey. 2003. Comparison of Method 1623 and Cell Culture-PCR for Detection of *Cryptosporidium* spp. in Source waters. *Appl. Environ. Microbiol.* 69(2),971-979

McTigue N.E., M. LeChevallier, H. Arora, and J. Clancy. 1998. *National Assessment of Particle Removal by Filtration.* Denver, CO: AwwaRF.

Nieminski E.C. and J.E. Ongerth. 1995. Removing *Giardia* and *Cryptosporidium* by Conventional Treatment and Direct Filtration. *Jour. Amer. Water Works. Assoc.* 87(9):96-106.

Nieminski E.C. 1994. Removing *Giardia* and *Cryptosporidium* Cysts through Direct Filtration and Conventional Treatment. *Proc. 1994 AWWA Ann. Conf.*, New York.

O'Melia C.R. and W. Ali. 1978. The role of retained particles in deep bed filtration. *Prog. Wat. Technol.* 10(5/6):167-182.

Ongerth J.E. 1990. Evaluation of Treatment for Removing *Giardia* Cysts. *Jour. Amer. Water Works. Assoc.* 82(6):85-96.

Ongerth J.E. and J.P. Pecoraro. 1995. Removing *Cryptosporidium* using Multimedia Filters. *Jour. Amer. Water Works. Assoc.* 87(12):83-89.

Patania N.L., J.G. Jacangelo, L. Cummings, A. Wilczak, K. Riley, and J. Oppenheimer. 1995. *Optimization of Filtration for Cyst Removal.* Denver, CO: AwwaRF.

Patwardhan V.S. and T. Chi. 1985. Sedimentation and Liquid Fluidization of Solid Particles of Different Sizes and Densities. *Chem. Eng. Sci.* 40(7):1051-1060.

Prasanthi H., S. Vigneswaran, H.B. Dharmappa. 1997. Effect of Particle Concentration on the Entire Cycle of Filtration. *Wat. Sci. Tech.* 35(8):91-102.

Smoluchowski M. 1917. Versuch einer mathematischen Theorie der Koagulations kinetic kolloider Losunger. *Zeitschrift Physicalische Chemie*. 92:129.

States S., K. Stadterman, L. Ammon, P. Vogel, J. Baldizar, D. Wright, L. Conley, and J. Sykora. 1997. Protozoa in River Water: Sources, Occurrence, and Treatment. *Jour. Amer. Water Works Assoc.* 89(9):74-83.

Thomas D.N., S.J. Judd, and N. Fawcett. 1999. Flocculation Modeling: A Review. *Water Research*. 33(7):1579-1592.

Tobiason J.E. and C.R. O'Melia. 1988. Physicochemical Aspects of Particle Removal in Depth Filtration. *Jour. Amer. Water Works Assoc.* 80(12):54-64.

US Environmental Protection Agency. 2003. National Primary Drinking Water Regulations: Long Term 2 Enhanced Surface Water Treatment Rule; Proposed Rule. *Federal Register*. 68(154):47639-47795.

US Environmental Protection Agency. 2006. National Primary Drinking Water Regulations: Long Term 2 Enhanced Surface Water Treatment Rule; Final Rule. *Federal Register*. 71(3):653-786.

Vigneswaran S. and J.S. Chang. 1989. Experimental testing of mathematical models describing the entire cycle of filtration. *Wat. Res.* 23(11):1413-1421.

Yao K.M., M.T. Habibian, and C.R. O'Melia. 1971. Water and Waste Water Filtration: Concepts and Applications. *Environ. Sci. Technol.* 5(11):1105-1112.

ABBREVIATIONS

APHA	American Public Health Association
AWWA	American Water Works Association
AwwaRF	Awwa Research Foundation
α	collision efficiency factor
C	concentration
C_0	influent concentration
cm	centimeter
CSCC	continuous separation channel centrifugation
°C	degrees Celsius
DABCO	diazabicyclo octane
d_C	collector diameter
DOC	dissolved organic carbon
η	collision frequency
f_A	fraction of labeled sample that was analyzed
FC	flow cytometry
f_F	number of fields per filter
FITC	fluorescein isothiocyanate
g	gravitational constant
gpd	gallon per day
gpd/ft^2	gallon per day per square foot
gpm	gallon per minute
gpm/ft^2	gallon per minute per square foot
HDPE	high-density polyethylene
L	liter
L	thickness of packed bed
L/mg/m	liter per milligram per meter
λ	filtration coefficient
mg	milligram
mg/L	milligram per liter
min	minute
mL	milliliter
µL	microliter
mL/min	milliliter per minute
mm	millimeter
µm	micrometer
MBR	Molecular Biology Resources

MCRF	Marshfield Clinic Research Foundation
N	concentration of pathogen in treated water
N_0	concentration of pathogen in source water
n_A	number of CSCC-concentrated aliquots
n_F	number of fields on the filter that were counted by the microscopist
n_P	number of pathogens that were counted by the microscopist
NIST	National Institute of Standards and Technology
nm	nanometer
NOM	natural organic matter
ntu	nephelometric turbidity unit
PAC	Project Advisory Committee
PBS	phosphate buffer solution
sec	second
SUVA	specific ultraviolet absorbance
TOC	total organic carbon
θ	packed bed porosity
UV254	ultraviolet absorbance at wavelength of 254 nm
UW	University of Wisconsin, Madison
U.S.	United States
USEPA	United States Environmental Protection Agency
UK	United Kingdom
V_A	volume of CSCC-concentrated aliquot that was analyzed
V_F	volume of CSCC-concentrated aliquot that was filtered for counting
V_L	volume of pellet that was labeled with FITC-conjugated antibody
V_P	volume of pellet achieved by swinging-bucket centrifuge
V_S	volume of sample collected
WSLH	Wisconsin State Laboratory of Hygiene

IWA Publishing's authorised EU representative for General Product Safety Regulations is Diane D'Arras, 15 rue Duret, 75116 Paris, France, e-mail: safety@iwap.co.uk.

Printed and bound by CPI Group (UK) Ltd, Croydon, CR0 4YY
17/09/2026
02218810-0018